The Audio Workstation Handbook

Series introduction

The Focal Press Music Technology Series is intended to fill a growing need for authoritative books to support college and university courses in music technology, sound recording, multimedia and their related fields. The books will also be of value to professionals already working in these areas and who want either to update their knowledge or to familiarise themselves with topics that have not been part of their mainstream occupations.

Information technology and digital systems are now widely used in the production of sound and in the composition of music for a wide range of end uses. Those working in these fields need to understand the principles of sound, musical acoustics, sound synthesis, digital audio, video and computer systems. This is a tall order, but people with this breadth of knowledge are increasingly sought after by employers. The series will explain the technology and techniques in a manner which is both readable and factually concise, avoiding the chattiness, informality and technical woolliness of many books on music technology. The authors are all experts in their fields and many come from teaching and research backgrounds.

Dr Francis Rumsey
Series Editor

Music Technology Titles

Acoustics and Psychoacoustics
David M. Howard and James Angus

The Audio Workstation Handbook
Francis Rumsey

**Computer Sound Synthesis for the Electronic Musician
(with CD-ROM)**
Eduardo Reck Miranda

Digital Audio CD and Resource Pack
Markus Erne
(Digital Audio CD also available separately)

Digital Sound Processing for Music and Multimedia
(with related website)
Ross Kirk and Andy Hunt

MIDI Systems and Control, Second Edition
Francis Rumsey

Sound and Recording: An Introduction, Third Edition
Francis Rumsey and Tim McCormick

Sound Synthesis and Sampling
Martin Russ

The Audio Workstation Handbook

Francis Rumsey

Focal Press

Focal Press
An imprint of Butterworth-Heinemann
Linacre House, Jordan Hill, Oxford OX2 8DP
225 Wildwood Avenue, Woburn, MA 01801–2041
A division of Reed Educational and Professional Publishing Ltd

℟ A member of the Reed Elsevier plc group

OXFORD   AUCKLAND   BOSTON
JOHANNESBURG   MELBOURNE   NEW DELHI

First published 1996
Reprinted 1999

British Library Cataloguing in Publication Data
A catalogue record for this book is available from the British Library.

Library of Congress Cataloguing in Publication Data
A catalogue record for this book is available from the Library of Congress.

ISBN 0 240 51450 5

Composition by Scribe Design, Gillingham, Kent
Printed and bound in Great Britain

Contents

Preface

The audio world is changing fast. When I wrote *Tapeless Sound Recording* six years ago it was the exception rather than the norm to record digital audio on computer hard disks, and most systems were run on dedicated hardware rather than desktop computers. Today it is becoming the norm rather than the exception to record and edit sound using computers and mass storage media, and the personal computer industry now produces off-the-shelf machines which are capable of storing and replaying real time audio and video with high quality.

The boundaries are no longer so clear between what we call professional audio and related fields such as multimedia production, the computer games market and the Internet. Professional audio engineers no longer have the sole claim to high quality sound, since even the most ordinary multimedia PC can now handle 'CD quality' audio. In some people's eyes the audio industry is getting consumed by the much larger computer and multimedia industry of which it is indeed a relation, but the truth is that sound still has very particular requirements and an identity of its own. Those working with multimedia need to have a detailed understanding of the audio technology they use, and those working in professional audio need a detailed understanding of related fields such as information technology, video and MIDI.

Many modern productions made using computer-based digital audio hardware and software fail to produce high quality sound because of a straightforward lack of knowledge about the factors which affect it. You may have a computer with 16 bit audio on board, but unless you understand something about digital audio you can still produce results with very poor technical quality. Democratisation of audio production has occurred because low cost, high quality equipment is available to everyone. Even so, rather as seen with desktop publishing, powerful hardware and software do not automatically make a skilled user, they simply give more people the *potential* to become skilled users.

I wrote this book principally to fill the gap left by two older books, *Digital Audio Operations* and *Tapeless Sound Recording*, which are coming to the end of their useful lives. I chose to call it *The Audio Workstation Handbook* to reflect what is now the key growth area in the audio field, and have included material on all of the main technologies associated with digital audio workstations. Areas of particular interest today are file interchange, networking, data reduction and digital video, and each of these has been incorporated. Alongside these is a substantial chapter on MIDI and sound synthesis for workstations, since much software integrates both real digital audio and MIDI processing, and multimedia workstations use MIDI controlled sound resources as a low overhead form of sound generation. I don't apologise for leaving out tape recording almost completely, although I acknowledge that it will remain important for many years to come. Many of the principles described in this book apply to systems other than audio workstations, making the book useful as a general reference on digital audio and associated fields.

Occasionally people have opined that the existence of the Internet, with its almost limitless wealth of free information, makes books such as this one unnecessary. 'If we can publish and receive all the latest information at our desktops', the story goes, 'why should we be concerned with books any longer, since they are out of date before they are published.' As anyone who has delved into Internet archives, web sites and mailing lists will know, they are full of useful practical tips and opinions, but someone has to perform the difficult and time-consuming task of digesting the information contained therein, deciding whether (a) it is accurate and (b) it is useful. Often it is neither. We live in a world where there is no end of information but little truth. I believe that carefully authored books will remain a valuable resource for the foreseeable future, with on-line electronic information sources fulfilling a complementary but distinctly different role.

The coverage in this book is slightly biased towards 'high-end' audio, and aims to educate people in those areas needed for the production of high quality sound. Having said that, I have put in considerable effort to make the book useful to those working with audio in other fields, and where possible have included descriptions of standards and approaches used in multimedia and desktop audio more broadly. Whilst an explanation of various standards is provided here, it is important to remember that this book is not a standards document. For specific details of the standards concerned the reader should refer to the appropriate publications, as described in the references and further reading recommendations. As usual I have tried to make the coverage as generic as possible, but I have taken examples from real products to illustrate concepts. Whilst the products may go out of date, the concepts illustrated will not. The reader will appreciate the impossibility of attempting to describe features and details of individual items of hardware and software in such a rapidly evolving field.

Francis Rumsey

1 Introduction to computer systems and terminology

The first chapter of this book is intended for those readers without a background in computing or information technology. It introduces a number of fundamental terms and principles relating to computers and digital systems which may be useful to people working with audio who have a mainly 'analog' background. It is not intended to be comprehensive, since the topics introduced in this chapter have filled many books in their own right. Those who feel comfortable with such material can ignore this chapter.

1.1 Digital workstations and the microcomputer

Digital workstations rely on microcomputers and digital signal processors (DSPs). The great revolution of recent years has been the creation of technology which can cost-effectively convert audio and video information into a digital form which can be easily manipulated using computers. Once in the digital domain, audio and video can be treated in many ways like other forms of data such as text and numbers.

As many will be aware, a microcomputer is not just a device with a screen and keyboard which sits on the desktop, used for typing letters and working out budgets; a microcomputer is fundamentally a system designed to manipulate binary numerical information (data), store the data, and communicate with peripheral devices in order to accept commands or give out

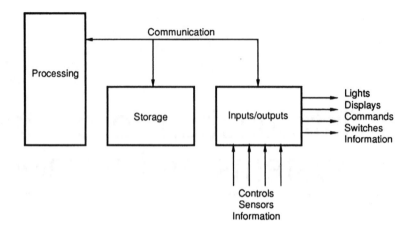

Figure 1.1 Conceptually, a microcomputer consists of inputs and outputs, processing and storage

information. The personal desktop computer takes in commands from a keyboard and other devices such as a mouse or trackball, is able to store it in various places such as in solid state memory or on a disk, and (by virtue of the software program running on the microprocessor) may perform operations on stored information in order to display a result or control a process.

A washing machine may contain a simple microcomputer designed to control a mechanical process. Its inputs will be from the front panel, and also from the various sensors built into the machine to tell it when the motors are rotating, which valves are open, whether the water pressure is high enough and so on. Its outputs will be to the display, and to the motors, relays and valves which initiate the various washing processes. It may store different washing programs for different purposes, and may process information coming from the sensors concerning the progress of the wash, the water temperature and so forth. Essentially therefore the microcomputer has inputs, outputs, processing and storage. The concepts are illustrated in Figure 1.1, and are expanded upon further in section 1.6.

The computer communicates with the outside world by means of interfaces or ports. These interfaces transmit and/or receive data, normally in electrical form (see section 1.6.4). The speed at which the data is passed through an interface depends on the application. In washing machine control there is normally no need for high speed communication of information between the various controls and the mechanism, since events take place relatively infrequently and are not time critical (at least not in the computing sense). Provided that the shirts stop spinning at roughly the right time the machine will work satisfactorily. With a digital audio workstation information may be arriving and leaving at a wide range of different rates. Information from the user interface will arrive relatively slowly, since even the fastest operator will only be generating a few commands per second. On the other hand, the interface connecting audio disk drives to the computer will be much faster, since real time digital audio information runs at a high data rate.

Once audio information and other control data is coded in a digital form and stored in a memory it can easily be changed. Effects which would either have been impossible with conventional analog techniques, or which would otherwise have taken a vast amount of time and equipment, become possible. Most people, for example, are familiar with the visual effects gymnastics performed on television, which are simply the result of taking the binary data representing the picture elements and rearranging them, modifying them, processing them using mathematical algorithms, and using them to create altered duplicates.

1.2 Analog and digital information

Before going any further it is necessary to compare analog and digital information. The human senses deal mainly with analog information, but computers deal internally with digital information, and thus there is the need for conversion between one domain and the other at various points.

Analog information is made up of a continuum of values, which at any instant may have any value between the limits of the system. For example, a rotating knob may have one of an infinite number of positions – it is therefore an analog controller (see Figure 1.2). A simple switch, on the other hand, can be considered as a digital controller, since it has only two positions – off or on. It cannot take any value in between. The brightness of light which we perceive with our eyes is analog information, and as the sun goes down the brightness falls gradually and smoothly, whereas a household light without a dimmer may be either on or off – its state is binary (that is it has only two possible states). A single item of binary information is called a bit (binary digit), and a bit can only have the value one or zero

Figure 1.2 (a) A continuously variable control such as a rotary knob is an analogue controller. (b) A two-way switch is a digital controller

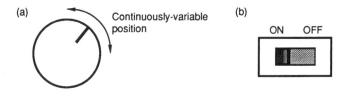

(a) Continuously-variable position

(b) ON   OFF

Figure 1.3 Electrical representation of analog and digital information. The rotary controller of Figure 1.2(a) could adjust a variable resistor, producing a voltage anywhere between the limits of 0 and +V, as shown in (a). The switch connected as shown in (b) allows the selection of either 0 or +V states at the output

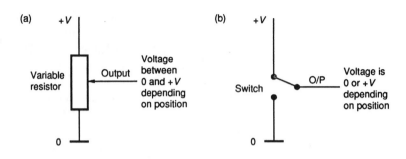

(corresponding, say, to high and low, or on and off states of the electrical signal).

Electrically, analog information may be represented as a varying voltage or current. If the rotary knob of Figure 1.2 is connected to a variable resistor and a voltage supply, its position will affect the output voltage (see Figure 1.3) which, like the knob's position, may have any value between the limits – in this case anywhere between zero volts and +V. The switch may be connected to a similar voltage supply, and in this case the output voltage can only be either zero volts or +V. In other words the electrical information which results is binary. The high (+V) state could be said to correspond to a binary one, and the low state to binary zero (although in many real cases it is actually the other way around). One switch can represent only one binary digit (or bit), but most digital information is made up of more than one bit, allowing digital representations of a number of fixed values.

Analog information in an electrical form is converted into a digital electrical form using a device known as an analog-to-digital (A/D) convertor – indeed it must be if it is to be handled by any logical system such as a computer. This process will be described in section 1.5. The output of an A/D convertor is a binary numerical value representing as accurately as possible the analog voltage which was converted.

Digital information made up of binary digits is inherently more resilient to noise and interference than analog information, as shown in Figure 1.4. If noise is added to an *analog* signal then it becomes very difficult to tell at any later stage in the signal chain what is the wanted signal and what is the unwanted noise, since there is no means of distinguishing between the two. If noise is added to a digital signal it *is* possible to extract the important information at a later stage, since it is known that only two states matter – the high and low, or one and zero states. By comparing the signal amplitude with a fixed decision point it is possible for a receiver to decide that everything above the

Figure 1.4 When noise is added to an analog signal, as shown at (a), it is not possible for a receiver to know what is the original signal and what is the unwanted noise. With the binary signal, as shown at (b), it is possible to extract the original information even when noise has been added. Everything above the decision level is high, and everything below it is low

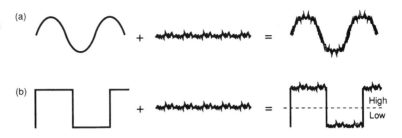

decision point is 'high', whilst everything below it is 'low'. Any levels in between can be classified in the nearest direction. Thus for any noise to influence the state of a digital signal, it must be at least large enough in amplitude to cause a high level to be interpreted as 'low', or vice versa.

The timing of digital signals may also be corrected to some extent, using a similar method, which gives digital signals another advantage over analog. If the timing of bits in a digital message becomes unstable, such as after having been passed over a long cable, resulting in timing 'jitter', the signal may be reclocked at a stable rate ensuring that the timing stability of the information is restored.

1.3 Binary number systems

1.3.1 Basic binary

In the decimal number system, each digit of a number represents a power of ten. In a binary system each digit or bit represents a power of two (see Figure 1.5). It is possible to calculate the decimal equivalent of a binary integer (whole number) by using the method shown. A number made up of more than one bit is called a binary 'word', and an 8 bit word is called a 'byte' (from 'by eight'). Four bits is called a 'nibble'. The more bits there are in a word the larger the number of states it can represent, with eight bits allowing 256 (2^8) states and sixteen bits allowing 65 536 (2^{16}). The bit with the lowest weight (2^0) is called the least significant bit or LSB, and that with the greatest weight is called the most significant bit or MSB. The term kilobyte or Kbyte is used to mean 1024 or 2^{10} bytes, and the term megabyte or Mbyte represents 1024 Kbytes.

Electrically it is possible to represent a binary word in either 'serial' or 'parallel' form. In serial communication only one connection need be used, and the word is clocked out one bit at a time using a device known as a shift register. The shift register

Figure 1.5 (a) A binary number (word or 'byte') consists of bits. (b) Each bit represents a power of two. (c) Binary numbers can be represented electrically in pulse-code modulation (PCM) by a string of high and low voltages

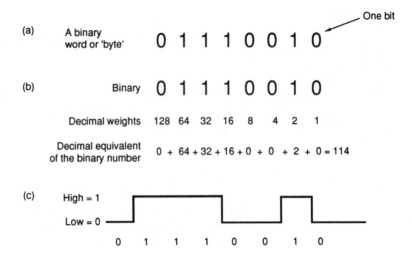

Figure 1.6 A shift register is used to convert a parallel binary word into a serial format. The clock is used to shift the bits one at a time out of the register, and its frequency determines the data rate. The data may be clocked out of the shift register either MSB or LSB first, depending on the device and its configuration

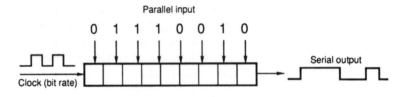

is previously loaded with the word in parallel form (see Figure 1.6). The rate at which the serial data is transferred depends on the speed of the clock. In parallel communication, each bit of the word is transferred over a separate connection.

Because binary numbers can become fairly unwieldy when they get long, various forms of shorthand are used to make them more manageable. The most common of these is hexadecimal. The hexadecimal system represents decimal values from 0 to 15 using the sixteen symbols 0–9 and A–F, according to Table 1.1, thus each hexadecimal digit corresponds to four bits or one nibble of the binary word. An example showing how a long binary word may be written in hexadecimal (hex) is shown in Figure 1.7 – it is simply a matter of breaking the word up into 4 bit chunks and converting each chunk to hex. Similarly, a hex word can be converted to binary by using the reverse process.

Figure 1.7 This 16 bit binary number may be represented in hexadecimal as shown, by breaking it up into 4 bit chunks (nibbles) and representing each chunk as a hex digit

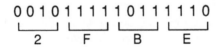

Table 1.1 Hexadecimal and decimal equivalents
to binary numbers

Binary	Hexadecimal	Decimal
0000	0	0
0001	1	1
0010	2	2
0011	3	3
0100	4	4
0101	5	5
0110	6	6
0111	7	7
1000	8	8
1001	9	9
1010	A	10
1011	B	11
1100	C	12
1101	D	13
1110	E	14
1111	F	15

1.3.2 Negative numbers

Negative integers are usually represented in a form known as 'two's complement'. Negative values are represented by taking the positive equivalent, inverting all the bits and adding a one. Thus to obtain the 4 bit binary equivalent of decimal minus five (-5_{10}) in binary two's complement form:

$$5_{10} = 0101_2$$
$$-5_{10} = 1010 + 0001 = 1011_2$$

Two's complement numbers have the advantage that the MSB represents the sign (1=negative, 0=positive), and that arithmetic may be performed on positive and negative numbers giving the correct result:

e.g. (in decimal):
$$5$$
$$+(-3)$$
$$=2$$

or (in binary):
$$0101$$
$$+ 1101$$
$$= 0010$$

The carry bit which may result from adding the two MSBs is ignored.

Figure 1.8 Negative numbers represented in two's complement form create a continuum of values where maximum positive wraps round to maximum negative, and bits change from all zeros to all ones at the zero crossing point

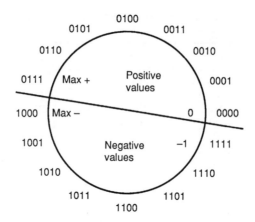

An example is shown in Figure 1.8 of 4 bit, two's complement numbers arranged in a circular fashion. It will be seen that the binary value changes from all zeros to all ones as it crosses the zero point, and that the maximum positive value is 0111 whilst the maximum negative value is 1000, so the values wrap around from maximum positive to maximum negative.

1.3.3 Fixed and floating point representation

Fixed point binary numbers are often used in digital audio systems to represent sample values. These are usually integer values represented by a number of bytes (2 bytes for 16 bit samples, 3 bytes for 24 bit samples, etc.). In some applications it is necessary to represent numbers with a very large range, or in a fractional form. Here floating point representation may be used. A typical floating point binary number might consist of 32 bits, arranged as four bytes, as shown in Figure 1.9. Three bytes are used to represent the *mantissa*, and one byte the *exponent* (although the choice of number of bits for the exponent and mantissa are open to variance depending on the application). The mantissa is the main part of the numerical value, and the exponent determines the power of two to which the mantissa must be raised. The MSB of the exponent is used to represent its sign, and the same for the mantissa.

Figure 1.9 An example of floating point number representation in a binary system

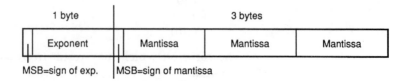

Figure 1.10 Symbols and truth tables for basic logic functions. The invertor shown on the right has an output which is always the opposite of the input. The circle on the invertor's output can be used to signify inversion on any input or output of a logic gate

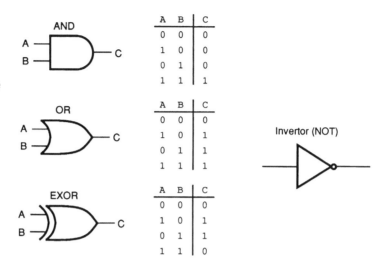

It is normally more straightforward to perform arithmetic processing operations on fixed point numbers than on floating point numbers, but signal processing devices are available in both forms.

1.4 Logical operations

Most of the apparently complicated processing operations that occur within a computer are actually just a fast sequence of simple logical operations. The apparent 'power' of the computer and its ability to perform complex tasks are really due to the speed with which simple operations are performed.

The basic family of logical operations is shown in Figure 1.10 in the form of a truth table next to the electrical symbol which represents each 'logic gate'. The AND operation gives an output only when both its inputs are true; the OR operation gives an output when either of its inputs are true; and the XOR (exclusive OR) gives an output only when one of its inputs is true. The invertor or NOT gate gives an output which is the opposite of its input, and this is often symbolised using a small circle on inputs or outputs of devices to indicate inversion.

1.5 Basic analog-to-digital and digital-to-analog conversion

It is not intended to cover this subject in detail here, but the basic principles will be given. Conversion is a very large subject in its own right, but enough information will be given here to enable the reader to understand what happens when an analog

9

control's position is digitised. The process of sound digitising is discussed in more detail in Chapter 2.

As already stated, an analog signal can have an infinite number of amplitudes, whereas a digital signal's amplitude can only have a certain number of fixed values. The number of fixed values possible with a digital signal depends on the number of bits in the binary words involved. In order to convert an analog signal into a digital signal it is necessary to measure its amplitude at specific points in time (called 'sampling'), and to assign a binary value to each measurement (called 'quantising'). The diagram in Figure 1.11 shows a rotary knob against a fixed scale running from 0 to 9. If one were to quantise the position of the knob it would be necessary to determine which point of the scale it was nearest, and unless the pointer was at exactly one of the increments the quantising process would involve a degree of error. It will be seen that the maximum error is actually plus or minus half of an increment, since once the pointer is more than halfway between one increment and the next it should be quantised to the next.

Quantising error is an inevitable side effect in the process of A/D conversion, and the degree of error depends on the quantising scale used. Considering binary quantisation, a 4 bit scale offers sixteen possible steps, an 8 bit scale offers 256 steps, and a 16 bit scale 65 536. The more bits, the more accurate the process of quantisation.

In older systems, the position of an analog control was first used to derive an analog voltage (as shown earlier in Figure 1.3), and then that voltage was converted into a digital value using an A/D convertor (see Figure 1.12). More recent controls may be in the form of binary encoders whose output is immediately digital. Unlike analog controls, switches do not need the services of an A/D convertor for their outputs to be usable by a computer, since a switch's output is normally binary in the first place. Only one bit is needed to represent the position of a simple switch.

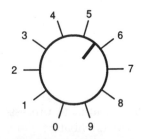

Figure 1.11 A rotary knob's position could be measured against a numbered scale such as the decimal scale shown. Quantising the knob's position would involve deciding which of the limited number of values (0–9) most closely represented the true position

Figure 1.12 In older equipment, a control's position was digitised by sampling and quantising an analog voltage derived from a variable resistor connected to the control knob

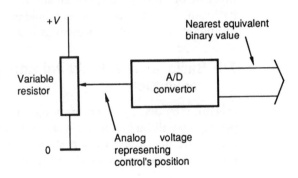

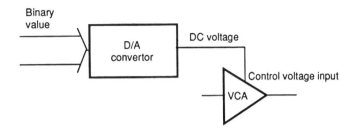

Figure 1.13 A D/A convertor could be used to convert a binary value representing a control's position into an analog voltage. This could then be used to alter the gain of a voltage-controlled amplifier (VCA)

The rate at which switches and analog controls are sampled depends very much on how important it is that they are updated regularly. Many analog mixing consoles sample the positions of automated controls once per television frame (40 ms in Europe), whereas some digital mixers sample controls as often as once per audio sample period (roughly 20 µs). Clearly the more regularly a control is sampled the more data will be produced, since there will be one binary value per sample.

Digital-to-analog conversion is the reverse process, and involves taking the binary value which represents one sample and converting it back into an electrical voltage. In a control system this voltage could then be used to alter the gain of a voltage-controlled amplifier (VCA), for example, as shown in Figure 1.13. Alternatively it may not be necessary to convert the word back to an analog voltage at all. Many systems are entirely digital and can use the binary value derived from a control's position as a multiplier in a digital signal processing operation. A signal processing operation may be designed to emulate an analog control process.

1.6 Basic computer system principles

This section contains an overview of the functions and operation of the key devices in a microcomputer system. Although practical implementations obviously differ considerably from each other, the principles remain much the same. Also, although computer systems become more complex all the time, it is still necessary to understand how a simple one works. The concepts involved in more complex computers are often just extensions of these basic ideas.

1.6.1 Buses

Figure 1.14 shows a basic block diagram of a microcomputer system, showing the main functional blocks. Before looking at specific devices within this system it is important to understand

Figure 1.14 Simple block diagram of a microcomputer (see text)

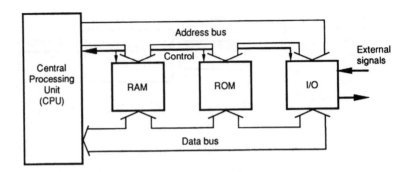

the function of a bus, which is normally a parallel collection of wires or printed circuit tracks each carrying one bit of a binary word. Thus an 8 bit data bus will carry data eight bits at a time in parallel, and a 16 bit address bus will carry sixteen address bits in parallel.

Information may be made to appear to travel in either direction on a bus, although direction is not really the right way of imagining it. The apparent direction of data flow simply depends on which device is placing the data on the bus and which device is reading it. The bus is normally shared between a number of devices, and data is routed between the central processing unit (CPU) and a device by 'enabling' the appropriate receiver or transmitter at the appropriate time. Clearly it is important to ensure that two devices do not attempt to place data on the bus at the same time.

1.6.2 The CPU

The central processing unit (CPU) is effectively the system controller, and its main functions are to sequence and interpret instructions which are fetched from memory, to perform logical operations on whole binary words, to store data temporarily and to monitor external requests for attention (interrupts). The CPU communicates with other devices using a data bus, an address bus and various control lines, as outlined in section 1.6.1. The CPU is connected to a crystal clock which generates a synchronisation signal at a rate of a number of megahertz. This clock drives the whole sequence of operations that occurs in the computer, because it is the changes of the clock signal which instigate the next event in the logical program sequence. Built into the CPU is a sequencer which determines the order of logical events in the many gates, stores and counters of the CPU. This sequencer is programmed by an instruction decoder which reads the binary words that form the instructions as they are fetched from memory. There is only a limited number of possible commands actionable by the CPU, and these commands are called the instruction set. On each cycle of the clock the sequencer steps one stage further through the sequence programmed by the last fetched instruction.

One of the most important devices within the CPU is the ALU (arithmetic and logic unit), which is a programmable device designed to perform logical/mathematical operations on the data. An ALU is really a large collection of gates of the type described above, and typically has two inputs which are a number of bits wide, and a number of control inputs and outputs. The control inputs are used to determine what logical operation will be performed, and 'carry in' and 'carry out' lines are used when the result of an operation is too large or too small and so overflows the MSB of the word. One of the inputs to the ALU is normally fed from a temporary store known as the accumulator, and the other input may be data read in over the data bus, for example.

The CPU contains a program counter which normally starts from zero at power up, and whose output is routed to the address bus in order to point to the location of the next command stored in the memory. When the clock has run the appropriate number of cycles (known as machine cycles) to run the sequencer through the current instruction, the program counter is incremented to the next address and the next instruction is fetched from memory. A typical sequence of events for a single instruction might go something like this:

> Place next instruction address onto address bus
> Read contents of that memory location into instruction decoder
> (Decode instruction to determine next step)
> Fetch next byte of data from memory
> Place that byte at one input to the ALU
> Add that byte to the byte contained in the accumulator
> Store the result in a temporary location
> Increment program counter

This is one instruction being executed, but it has taken a number of cycles of the clock.

The CPU also contains a temporary store called the stack, which is configured in the last-in–first-out mode. The stack is like a sprung plate holder, such as might be found in a cafeteria, which holds a pile of plates. The last plate to be pushed on to the top of the pile is the first one to come off the top if someone wants a plate. Data is sometimes pushed temporarily onto the stack, using it as a holding place while another action is executed, whereafter the data is pulled back off the stack.

The instruction set of a typical CPU contains commands such as those to move data from one location to another, those to perform mathematical operations on a pair of numbers and those to jump the program execution to a new memory address. There are also important commands to read and write data to input/output (I/O) ports, which is the means by which data is communicated to and from the outside world. This is covered further in section 1.6.4.

1.6.3 ROM and RAM

There are two main types of solid-state memory in the typical computer: read only memory (ROM), and random access memory (RAM). (Solid-state devices are normally based on small slivers or 'chips' of silicon which have been locally 'doped' to behave as very large collections of electronic devices such as transistors, resistors and capacitors. Tiny wires are attached to discrete points on the silicon wafer, and these are connected to the pins on the chip packaging so as to allow connection to the outside world.)

There are many sub-types of ROM and RAM with subtle differences. All memory devices store particular bit patterns in different locations, known as addresses. These may be imagined as a set of 'pigeonholes', each with a unique matrix reference, as illustrated in Figure 1.15. These memory addresses may hold data that is anything from one bit wide to many. In the case of 1 bit memory chips it may be necessary to configure them in an

Figure 1.15 Memory addresses may be likened to the 'grid references' of location in this matrix. The shaded square has the address F6

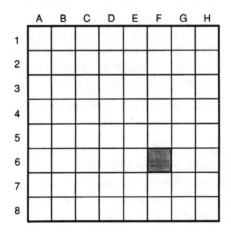

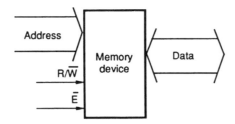

Figure 1.16 A memory device connected to data and address buses. The R/W line determines the direction of data flow (high for read, low for write). The enable line enables this particular device to read or write

array in order to store 8 bit words. Memory is characterised in block diagram form as shown in Figure 1.16, showing an address bus, a data bus, an enable line and a read/write (R/W) line (ROM has no R/W line because it is read-only). In order to write a byte of data to the memory, an address is presented on the address bus and the appropriate data is presented on the data bus. The R/W line is changed to the write state, and the memory chip concerned is enabled by holding the enable line in the appropriate state. Read operations are performed in a similar manner, except with the R/W line in the opposite state.

ROM is designed for data which will not be changed by the system, such as parts of the operating system software (the commands which program the CPU). True ROM is programmed by the manufacturer by blowing tiny fusible links in the chip at certain memory addresses in order to set particular bit patterns. These are then permanently stored and cannot be erased, thus making true ROM a good permanent store for limited amounts of data (up to a few megabytes). The data is retained even with no power to the chip. Other forms of ROM exist which may be modified with varying degrees of ease. EPROM is erasable programmable ROM, and may be erased by exposing a small window to ultraviolet light for around 20 minutes, after which it may be reprogrammed in a so-called 'PROM blower' by applying a higher than normal voltage to the pins of the EPROM. EEPROM (electrically erasable PROM) and EAROM (electrically alterable ROM) are further variations, and these may be reprogrammed *in situ* by the system itself, although not normally as easily as RAM may be programmed. Such devices are useful in systems where a semi-permanent form of storage is required, but where the data may need to be modified on occasions.

RAM is capable of being both written to and read from. It is a very fast form of temporary store, in that data stored in it can be accessed typically in under 100 nanoseconds, but the store loses its data when the power is turned off (unless some form of battery backup has been arranged). Desktop computers often use a form of RAM called dynamic RAM (DRAM) whereby the

15

memory devices are arranged on small plug-in modules called SIMMs (structured in-line memory modules), capable of storing many megabytes of data each. The amount of RAM available on a single chip increases every year, and thus it is difficult to be precise about storage capacity. RAM is used in computer systems for storing the temporary data which results from inputs and program operations, but also may be used for operating system software which has been temporarily uploaded from a more permanent external store such as a disk drive. It is important to distinguish between the RAM in a computer and its disk storage capacity, since they are both often referred to as 'memory'.

1.6.4 I/O ports

The I/O (input/output) ports are the third main element of a basic microcomputer system. As mentioned earlier, these are the system's windows on the outside world, without which the piece of equipment would be virtually useless. Some of these ports may appear as physical connectors on the rear panel of the equipment concerned, and others may be connected to internal peripherals such as disk drives, in which case the ports may not be available for general purpose use. One very common use of I/O ports in some desktop computers is to route data to a video card, so that images can be displayed. I/O ports may be either parallel or serial.

The important point to raise here is the way in which information is communicated to and from the ports, because there are a variety of ways of doing this, some of which will be outlined here. First of all, it will be seen from the main block diagram that the I/O ports in a simple computer are generally connected to the same data and address bus structure as the memory, but in some systems there is a separate control line from the CPU to indicate whether it is addressing memory or I/O – they are not both enabled at the same time. An alternative is to use so-called 'memory-mapped I/O', in which I/O ports occupy memory addresses. Each I/O port has at least one address, and usually a number, corresponding to different aspects of the port's function. One such address may be the port's 'control register' which allows the CPU to program such aspects as the clock speed of a serial interface. By writing to an I/O address the CPU can transfer a byte of data from a temporary internal store to a peripheral device.

Similarly, by reading a particular control register address it is possible to tell whether a byte of data has been received by a port. This is called 'polling', and is the CPU's means of determining whether there is any data to be had from any of the ports. It is rather like a teacher asking each member of the class in turn whether they want to say something – eventually the teacher may come across one who does, but if only two out of thirty have something to say it is a fairly time consuming and wasteful exercise. An alternative is for the person who has something to say to raise a hand and wait for the teacher to respond. In computer terms this is called 'interrupting', and is often used as a means of flagging-up the presence of new data at a port.

When a port receives a byte of data it may be able to raise the interrupt flag, which is one of the control lines connected to the CPU. The CPU then finishes the current task and acknowledges the interrupt using another line (IACK), after which a number of possible scenarios may occur. Either the CPU must then address each port in turn to find out which one interrupted, or, more efficiently, on receipt of the IACK flag, the port which generated the interrupt may place its own address or another unique identifier onto the data bus, which may then be read by the CPU to determine the source of the interrupt. This data value can be used to cause the CPU to jump its program counter to a new memory address which contains the start of a small subroutine to handle that particular interrupt, which in a simple case might involve reading the byte of data from the input buffer of the port and storing it in a RAM location.

When a lot of interrupts are likely to occur, such as when there are a large number of ports, it may be necessary to prioritise them, and often a dedicated device is used for this purpose which is capable of taking in a large number of interrupts and arranging for the CPU to service those with the highest priority first. Those which are time critical, such as synchronisation messages, will clearly have to be serviced before those which are not. Ports may need a small amount of buffer memory associated with them in order to allow data to be stored temporarily, in case the CPU is not able to service an interrupt immediately. If another byte of data were to arrive at the same port before the first one had been collected it would be lost unless there were some temporary storage.

Commonly encountered general purpose I/O ports on desktop computers are serial ports conforming to RS232 or RS422 standards, parallel ports such as the Centronics printer interface, and fast parallel ports such as SCSI (the Small Computer Systems Interface), which is often used for connecting hard disk drives.

1.6.5 *Factors affecting computer system performance*

Aside from the question of software design, there are a number of hardware issues that affect the performance of any system based around a microprocessor. It is useful to know about these because it may help when deciding what equipment to use for a certain purpose, and whether one device will perform better than another.

The main thing that most people are interested in when comparing computers is the speed with which they will perform certain operations. This is important because a faster computer will be capable of performing complex tasks without requiring the user to go away and make a cup of coffee while the computer works on the problem. A fast computer may be able to perform certain real-time operations that a slower machine might not, because real-time tasks such as animation, video, digital audio recording and music sequencing all require that a certain number of operations are completed in a specific time frame.

Clock speed is one of the factors dictating performance, since, as already explained, this dictates the rate at which instructions are sequenced. Early computers used in MIDI (Musical Instrument Digital Interface – see Chapter 8) equipment tended only to be capable of clock speeds in the region of 2 MHz, which is slow by modern standards. An older desktop computer found today might be running at around 8 MHz, and the super-fast machines at rates of over 100 MHz. Clock speeds of CPUs are increasing all the time, and so one should expect continued improvements in this area. It is not possible to say what minimum clock speed is required for a particular task, since it is only one of the deter-mining factors.

Bus width is a second factor. In general, the wider the bus the faster the machine, since a wide bus makes it possible to trans-fer more data between devices per instruction. For example, a 4 byte floating point number (that is one with a number of decimal places and an exponent, or power of ten) would take four fetches to load it into the CPU with an 8 bit data bus, but only one with a 32 bit bus. It is very important, though, to distinguish between the width of the *system's* data bus and that within the CPU. The width of the CPU's internal data bus dictates what it can manip-ulate and store internally, and the external bus dictates how much of this data can be fed to and from the peripherals and memory in one go. Some so-called 32 bit CPUs have only 16 bit external data buses, whereas full 32 bit CPUs also have 32 bit external buses.

The speed of RAM installed in a system may also affect the apparent speed of operation, since slower RAM requires that the CPU waits a certain number of machine cycles for it to produce stable data after having been addressed (known as 'wait states'). Faster RAM allows the use of no wait states, thus speeding up access to stored data during program execution.

A particular type of processor, known as the RISC processor, is increasingly used in digital equipment because it operates using a reduced (simplified) instruction set in order to be able to run at very high speeds. For a given clock speed the RISC-based machine will tend to carry out more operations per second than a conventional microprocessor-based machine.

There are also a number of miscellaneous factors which will affect the performance of a computer-based device. These include whether or not any co-processors are installed, and whether any form of fast memory cacheing is used. Co-processors are additional CPUs designed to share or remove some of the processing load, and these often deal with such functions as mathematical operations or graphics processing. Whether a co-processor will give any improvement in speed depends largely on the task in hand, and whether the software is designed to benefit from co-processing. The same is true of fast memory cacheing, which is a means of storing the most recently used data in a small amount of very high speed RAM, either close to or installed in the CPU, such that it can be accessed more quickly and easily than general purpose RAM.

1.7 Mass storage

ROM and RAM are both solid-state forms of storage, and their advantage is that they can be built into the electrical bus structure of the computer and accessed very quickly (within tens or hundreds of nanoseconds), but there is often also the need for storage which can be removed or which can be used to keep much larger amounts of data than could reasonably or economically be stored in a solid-state form. The larger, more permanent forms of storage are called mass storage devices, and usually take the form of disk drives or tape drives. Using such devices it is possible to store an almost unlimited amount of data, especially if removable media are used. The access time to data stored on such peripherals tends to be quite a lot longer than for internal RAM, being of the order of a few milliseconds for the typical hard disk drive, and a number of seconds for tape drives.

Data stored on mass storage devices is formed into 'files', and a catalogue of file locations is kept in an index known as a 'directory'. Mass storage is often connected to computer I/O ports using the SCSI, which is a fast parallel interface commonly found on desktop computers. Mass storage media are discussed in greater detail in Chapter 4.

1.8 Digital signal processors (DSPs)

Digital signal processors (DSPs) are increasingly used where the processing operations to be carried out are intensively mathematical. A conventional CPU can perform mathematical functions such as multiplication, but compared to a DSP it is normally slow. This is because the DSP's internal architecture is optimised for the sole purpose of performing lots of numerical operations at high speed, whereas the typical CPU is a general purpose device.

DSPs feature on many purpose-built expansion cards for multimedia PCs, such as those for audio and video processing, since many of the operations carried out on audio and video require numerical manipulation. Tonal equalisation and the addition of artificial reverberation are just two examples of DSP operations common with audio data. Data reduction processes, such as MPEG and JPEG (discussed in sections 3.5, 3.6, 3.7 and 7.3), also require high-powered DSP to be able to operate in real time.

DSP devices are found in both fixed and floating point types, the former acting only on integer data.

1.9 The role of software in computer systems

Software has been mentioned a number of times in the preceding sections of this chapter, and it will be mentioned many times again, but what is it and what is its role in the computer system? This section will give an introduction to the two main areas of software which most people will come across – operating systems and applications software.

1.9.1 What is software?

The CPU of a computer is a logical machine which can step through a preprogrammed sequence of operations at a speed defined by the clock. It has no magical 'knowledge' or 'skill' and it relies on being told what to do next. Depending on data presented to the CPU at different stages in its machine cycle, the sequence of operations may be changed. This is really the role of software – to program the CPU and to provide it with some of the data to be operated upon by the program instructions (the rest comes from the outside world, or is the result of operations performed during the execution of the program).

Within the CPU is an instruction decoder (see section 1.6.2) which reads a byte of data (software) supplied from the memory and sets a complex collection of logical gates, latches and registers to a particular state, so that on the next cycle of the clock a particular collection of logical changes will occur. It is an electronic version of the mechanical adding machine where the user sets up certain mechanical conditions within a device consisting of cogs, ratchets and levers, after which the turning of a handle (like the cycling of the clock) executes a mechanical sequence which depends on the initial mechanical state of the components.

Thus software is a sequence of binary instructions and data which causes the CPU to act in a particular way as the clock cycles. Software at this level is called 'machine code', and the instructions are called 'low level' instructions, since they are at the raw binary level which the CPU can interpret directly. It is this data which is actually stored in the computer's memory and which is used by the CPU, but unfortunately to write a program at this level is exceedingly tedious.

What is required is a form of 'high level' language, and there are many of these with different strengths. One up from machine code is 'assembler' which is really a collection of mnemonics representing machine code instructions which are easier to use than pure binary codes. Writing a program in assembler is very tedious but quite efficient from a running point of view, because it is written at the 'nuts and bolts' level of the system. At the highest level are programming languages like 'Hypertalk' that look almost like speech, containing lines like:

put the contents of field 1 into line 3 of field 5

and these languages are designed for use by people who want to write their own software but do not wish to get too deeply into the inner workings of the computer. The drawback of many such languages is that the programs run more slowly because a lot of interpretation has to take place to turn these 'English' commands into machine code before the operation can be executed.

1.9.2 Compiled and interpreted programs

There are two basic approaches to turning high level programs into machine code – compiling and interpreting. A compiler is a software program which takes the whole of a high level program and turns it into efficient machine code before it can be run. Execution is then a matter of running the machine code or compiled version of the software. This approach is tedious when writing the program in the first place – because the software has to be compiled each time before it can be run – but it is used widely because the compiled program then runs very quickly. An interpreter codes the program lines into machine code while the program is running, and thus it is convenient to use while programming but very slow in execution due to the need for interpreter intervention all the time.

The two approaches are very similar to language translation for speech. If you talk to a foreign person via an interpreter it is quite convenient but you have to keep stopping every line for the interpreter to translate what you said. If what you wanted to say was written down on paper and translated (compiled) first, then either you or a native speaker could use it directly and more quickly, but it would require an initial delay for the translation.

1.9.3 Object-orientated software

Object-orientated software is a concept which has grown considerably in importance over the last few years, and it presents another level of high level interaction with the computer. In an object-orientated environment, software 'modules' are written, each of which performs a particular function, and these are termed software 'objects'. An object may have a number of inputs and outputs. The objects may then be combined and interconnected, often graphically, in various ways such that the output of one object feeds another's input, and so on, creating what is really a virtual machine with controls, processors, inputs, outputs and displays. The object-orientated concept extends into corresponding programming languages as well.

1.9.4 Operating systems

An operating system (OS) is a software program which runs 'in the background' all the time that a computer-controlled device is turned on. It is this which gives a particular computer system its peculiar characteristics, such as how it displays information, what commands it accepts, how disks are formatted and how memory is organised. It is also likely to be dedicated to a particular family of CPU, because the operating system must issue low level instructions in CPU-specific code. The operating system is

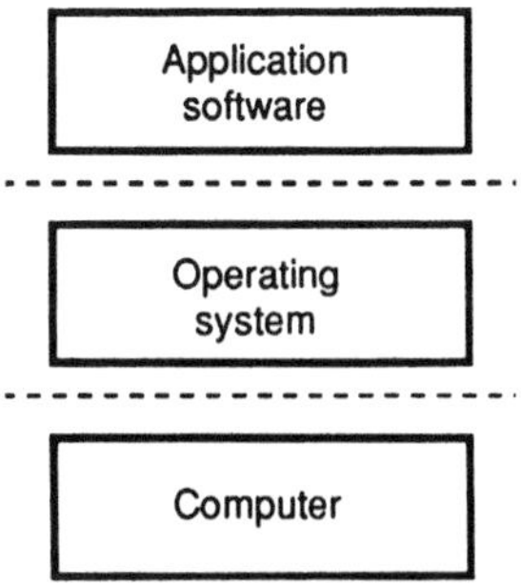

```
┌─────────────────────┐
│    Application      │
│    software        │
└─────────────────────┘
- - - - - - - - - - - -
┌─────────────────────┐
│    Operating       │
│    system          │
└─────────────────────┘
- - - - - - - - - - - -
┌─────────────────────┐
│    Computer        │
└─────────────────────┘
```

Figure 1.17 The operating system forms a layer between the application software program and the inner workings of the computer

a fundamental 'toolkit' which is called upon by higher level applications to perform the more mundane tasks such as disk storage, handling I/O from the keyboard, writing to the display, and so on. It gives a degree of consistency to the operation of a particular system, and avoids the need for programmers to write very basic functions such as those dealing with mathematical operations every time they write an application. The operating system is a layer of software intervention which resides in between the application and the microprocessor, as shown in Figure 1.17. It is still possible for applications to deal directly with the CPU, but most tasks can be passed via the operating system.

Examples of common personal computer operating systems are Microsoft's MS-DOS (designed for the 80x86 CPU family) and Apple's System 7 (designed for the 680xx series of CPUs). Larger computers and mainframes often operate under the Unix system. Applications written for these operating systems must conform to certain basic guidelines which describe how the program should interact with the operating system. Older operating systems tended to be text based, in that the only way that a user could interact with the OS was by typing strings of often unrememberable commands from a QWERTY keyboard, every character and space of which had to be correct otherwise the user would be presented with an equally incomprehensible error message. Recent systems have been designed to be more 'user friendly', often employing graphical user interfaces, as discussed below.

1.9.5 Graphical user interfaces

Graphical user interfaces, or GUIs as they are sometimes known, do not use text as the main means of communication with the operating system. Instead a whole graphical 'layer' is placed between the user and the operating system such that the user may only have to point at what he wants and select it. Often a mouse or other type of pointing device is interfaced to the computer via a port, to control a displayed arrow on the screen which moves as the mouse moves. A button on the mouse is then used to select whatever is being pointed at. Instead of the user having to remember commands precisely, a choice of options is presented in menu form and the user selects whichever is required. Such environments have been called WIMP (Windows, Icons, Mouse and Pointer) environments.

GUIs have done much to bring the power of the desktop computer to those who don't feel 'computer-minded', although dedicated computer 'boffins' often profess a dislike for them because they tend to slow down the operating system. This effect on speed is true, but the benefits far outweigh the disadvantages, and the dislike of the GUI by such people is probably more to do with a wish to continue the pretence that computers are only for people who understand them!

A GUI such as 'Windows' was designed as a layer on top of an existing operating system (MS-DOS), in an attempt to make it more user friendly. The result is rather slower than a GUI which was designed as an integral feature of the OS from the start, such as Apple's System software. It is likely that GUIs of some sort will be a key feature of computer OS's for the foreseeable future, possibly coupled with elements of voice control and handwriting recognition as means of providing alternative commands.

1.9.6 Filing structures

In a text-based OS such as MS-DOS, files on a disk are grouped into 'directories', which are virtual catalogues of subsets of the disk contents. Directories are a useful way of subdividing the disk contents in order that a track can be kept of stored data, and projects can be grouped together. An almost infinite number of subdirectories is possible, dividing the disk contents in a type of tree structure, as shown in Figure 4.18. A GUI may present this tree structure as a collection of nested folders, within which can be files or more folders. The text-based OS will allow the user to define a 'path' to the required file by typing in a text string representing the directories concerned and the file name, e.g.:

c:songs\song3.mid

whereas a GUI will allow this to be done using a means of opening up folders or selecting the appropriate subdirectory from a menu.

File names may be limited to a certain number of characters, and may allow the addition of an extension, often of three letters, to define the file as of a particular type. They may also forbid the use of certain characters such as spaces. GUI OS's such as the Macintosh allow long filenames (up to 32 characters) with almost no restriction on what characters can be used, allowing files to be given more meaningful names such as '1993 Accounts (2nd draft)'.

1.9.7 Operating system extensions

An extension is a piece of software which adds to the OS's features or characteristics, and is often loaded from disk when the system starts up. It resides in memory and operates in the background, just as if it were part of the OS. An example of a system extension is Apple's QuickTime software which adds the capability for real-time replay of audio and video to a basic Macintosh computer. It is quite common for extensions to cause problems in computer systems, since they may interfere with the normal mode of operation and cause some applications to function incorrectly or even fail altogether, particularly if the extension does not conform fully to the OS guidelines. This is normally a matter for experimentation, and there are a number of utilities which exist to manage the turning on and off of system extensions.

1.9.8 Multitasking

A true multitasking OS is one which allows a number of applications to reside in different areas of RAM and run concurrently. The number of applications which may do this is normally limited by the amount of RAM, although a technique known as 'virtual memory' makes it possible to treat a proportion of disk capacity as if it were (slower) RAM. Multitasking can be useful when, for example, one application is performing a tedious task and the user wishes to work on another while the tedious task progresses. Multitasking requires a fast computer because it is required to perform more than one job at a time. When heavy processing tasks are taking place in the background, the foreground application can run much more slowly, and on slower machines this can make the foreground application virtually unusable.

A real-time multitasking system is one in which time is shared between applications in such a way that actions are completed within certain defined time limits. A number of desktop operating systems are not truly multitasking in this respect. Multitasking should be distinguished from the simpler 'program switching' arrangement in which a number of applications reside concurrently in RAM and may be switched between. With program switching, inactive applications do not normally continue to operate in the background, although it is still a useful thing to have since it saves closing one job and loading another application from disk every time you wish to change between them.

2 Digital audio

The purpose of this chapter is to give an introduction to those aspects of the theory and practice of digital audio which are important for workstations. For this reason the topic of error correction features only very briefly, since systems which rely on computer mass storage media do not normally need purpose-designed audio error correction. The chapter concentrates on those issues which affect sound quality in digital audio systems, and introduces the topic of audio DSP. Although when implemented and used correctly, the sound quality of digital audio can be very high, there are many pitfalls for the unwary and many ways in which quality can be compromised. The user who understands a little of the theory will normally be able to extract the most from a system, whatever its limitations.

The block diagram in Figure 2.1 shows the generalised construction of a digital audio system. Analog audio signals are converted into binary values during A/D conversion. These

Figure 2.1 Generalised block diagram of a digital audio system

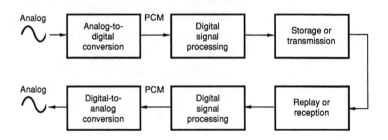

binary values may be modified in a variety of fashions using digital signal processing (DSP), after which they may be coded and stored or transmitted. Further processing may be applied after storage or transmission, followed by conversion back to the analog domain.

2.1 A/D conversion

The process of A/D conversion is of paramount importance in determining the ultimate quality of the digital audio signal. Indeed the quality of the audio, once converted, can never be made any better, only worse. Some applications deal with audio purely in the digital domain, in which case A/D conversion is not an issue, but most operations require the acquisition of audio material from the analog world at one time or another. The quality of convertors varies very widely in digital audio workstations and their peripherals because the price range of such workstations is also great. Some stand-alone professional convertors can easily cost as much as the complete digital audio hardware and software for a desktop computer. One can find 16 bit convertors built in to many multimedia desktop computers now, but these are really rather low performance devices when compared with the best available. As will be seen below, the sampling rate and the number of bits per sample are the main determinants of the quality of a digital audio signal, but the quality of the convertors determines how closely the sound approaches those limits.

Despite the above, it must be admitted that to many people one 16 bit convertor sounds very much like another, and there is a law of seriously diminishing returns when one compares the increased cost of good convertors with the perceivable improvement in quality. Convertors are very much like wine in this respect.

2.1.1 Sampling

An analog audio signal is a time-continuous electrical waveform, and the analog-to-digital convertor's task is to turn this signal into a time-discrete sequence of binary numbers. The sampling process employed in an A/D convertor involves the measurement or 'sampling' of the amplitude of the audio waveform at regular intervals in time (see Figure 2.2). From this diagram it will be clear that the sample pulses represent the instantaneous amplitudes of the signal at each point in time. The samples can be considered as like instantaneous 'still frames' of the audio

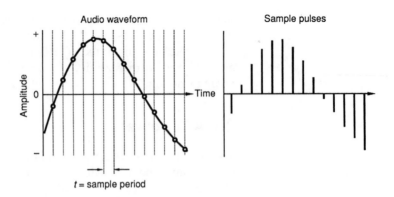

Figure 2.2 An arbitrary audio signal is sampled at regular intervals of time t to create short sample pulses whose amplitudes represent the instantaneous amplitude of the audio signal at each point in time

signal which together and in sequence form a representation of the continuous waveform, rather as the still frames which make up a movie film give the impression of a continuously moving picture when played in quick succession.

In order to represent the fine detail of the signal it is necessary to take a large number of these samples per second, and the mathematical sampling theorem proposed by Shannon[1] indicates that at least two samples must be taken per audio cycle if the

Figure 2.3 In the upper example many samples are taken per cycle of the wave. In the lower example less than two samples are taken per cycle and thus it is possible for another lower-frequency wave to be reconstructed from the samples. This is one way of viewing the problem of aliasing

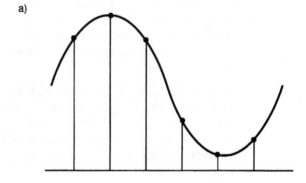

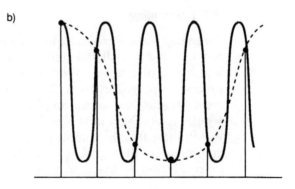

Figure 2.4 In pulse amplitude modulation, the instantaneous amplitude of the sample pulses is modulated by the audio signal amplitude (positive only values shown)

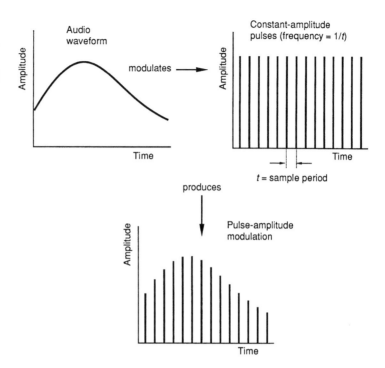

necessary information about the signal is to be conveyed. It can be seen from Figure 2.3 that if too few samples are taken per cycle of the audio signal then the samples may be interpreted as representing a wave other than that originally sampled, and this is one way of understanding the phenomenon known as aliasing. An 'alias' is an unwanted product of the original signal which arises when the sampled signal is reconstructed during D/A conversion.

Another way of visualising the sampling process is to consider it in terms of modulation, as shown in Figure 2.4. The continuous audio waveform is used to modulate a regular chain of pulses. The frequency of these pulses is the sampling frequency. Before modulation all these pulses have the same amplitude (height), but after modulation the amplitude of the pulses is modified according to the instantaneous amplitude of the audio signal at that point in time. This process is pulse amplitude modulation (PAM). The frequency spectrum of the modulated signal is as shown in Figure 2.5. It will be seen that in addition to the 'baseband' audio signal (the original spectrum before sampling) there are now a number of additional spectra, each centred on multiples of the sampling frequency. Sidebands have been produced either side of the sampling frequency and its multiples, as a result of the amplitude modulation, and these

Figure 2.5 The frequency spectrum of a PAM signal consists of a number of repetitions of the audio baseband signal reflected on either side of multiples of the sampling frequency

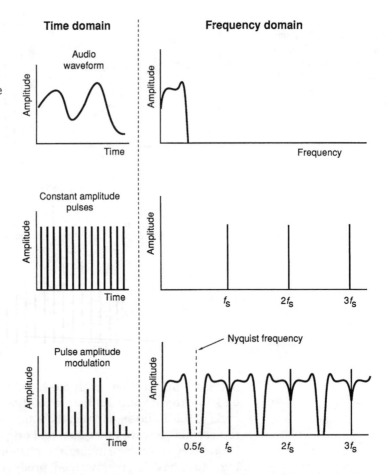

extend above and below the sampling frequency and its multiples to the extent of the base bandwidth. In other words these sidebands are pairs of mirror images of the audio band.

2.1.2 Filtering and aliasing

It is relatively easy to see why the sampling frequency must be at least twice the highest baseband audio frequency from Figure 2.6. It can be seen that an extension of the baseband above the Nyquist frequency results in the lower sideband of the first spectral repetition overlapping the upper end of the baseband. Two further examples are shown to illustrate the point – the first in which a baseband tone has a low enough frequency for the sampled sidebands to lie above the audio frequency range, and the second in which a much higher frequency tone causes the lower sampled sideband to fall well within the baseband, forming an alias of the original tone.

Figure 2.6 Aliasing viewed in the frequency domain. In (a) the audio baseband extends up to half the sampling frequency (the Nyquist frequency f_n) and no aliasing occurs. In (b) the audio baseband extends above the Nyquist frequency and consequently overlaps the lower sideband of the first spectral repetition, giving rise to aliased components in the shaded region. In (c) a tone at 1 kHz is sampled at a sampling frequency of 30 kHz, creating sidebands at 29 and 31 kHz (and at 59 and 61 kHz, etc.). These are well above the normal audio frequency range, and will not be audible. In (d) a tone at 17 kHz is sampled at 30 kHz, putting the first lower sideband at 13 kHz – well within the normal audio range. The 13 kHz sideband is said to be an alias of the original wave

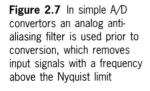

The aliasing phenomenon can be seen in the case of the well-known 'spoked-wheel' effect on films, since moving pictures are also an example of a sampled signal. In film, still pictures (image samples) are normally taken at a rate of 24 per second. If a rotating wheel with a marker on it is filmed it will appear to move round in a forward direction as long as the rate of rotation is much slower than the rate of the still photographs, but as its rotation rate increases it will appear to slow down, stop, and then appear to start moving backwards. The virtual impression of backwards motion gets faster as the rate of rotation of the wheel gets faster, and this backwards motion is the aliased result of sampling at too low a rate. Clearly the wheel is not really rotating backwards, it just appears to be.

If audio signals are allowed to alias in digital recording one hears the audible equivalent of the backwards-rotating wheel

Figure 2.7 In simple A/D convertors an analog anti-aliasing filter is used prior to conversion, which removes input signals with a frequency above the Nyquist limit

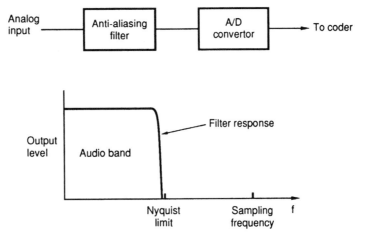

– that is, sound components in the audible spectrum which were not there in the first place, moving downwards in frequency as the original frequency of the signal increases. In basic convertors, therefore, it is necessary to filter the baseband audio signal before the sampling process, as shown in Figure 2.7, so as to remove any components having a frequency higher than half the sampling frequency (known as the Nyquist frequency).

In real systems, and because filters are not perfect, the sampling frequency is made slightly higher than twice the highest audio frequency to be represented, allowing for the filter to roll off slightly more gently. The filters incorporated into both D/A and A/D convertors have a pronounced effect on sound quality, since they determine the linearity of the frequency response within the audio band, the slope with which it rolls off at high frequency, and the phase linearity of the system. In a non-oversampling convertor, the filter must reject all signals above half the sampling frequency with an attenuation of at least 80 dB. Steep filters tend to have erratic phase response at high frequencies, and may exhibit 'ringing' due to the high 'Q' of the filter. Steep filters also have the added disadvantage that they are complicated to produce. Although filter effects are unavoidable to some extent, manufacturers have made considerable improvements to analog anti-aliasing and reconstruction filters, and these may be retro-fitted to many existing systems with poor filters. A positive effect is normally noticed on sound quality.

The process of oversampling (see below) has helped to ease the problems of analog filtering, since it shifts the first repetition of the baseband up to a much higher frequency, allowing the use of a shallower filter.

2.1.3 Quantisation

After sampling, the modulated pulse chain is quantised. In quantising a sampled signal the range of sample amplitudes is mapped onto a scale of stepped values, as shown in Figure 2.8. The quantiser determines which of a fixed number of quantising intervals (of size Q) each sample lies within, and then assigns it a value which represents the mid-point of that interval. This is done in order that each sample amplitude can be represented by a unique binary number in pulse code modulation (PCM). In linear quantising each quantising step represents an equal increment of signal voltage, and in a binary system the number of

Figure 2.8 When a signal is quantised, each sample is mapped to the closest quantising interval Q , and given the binary value assigned to that interval. (Example of a 3 bit quantiser shown). On D/A conversion each binary value is assumed to represent the voltage at the mid-point of the quantising interval

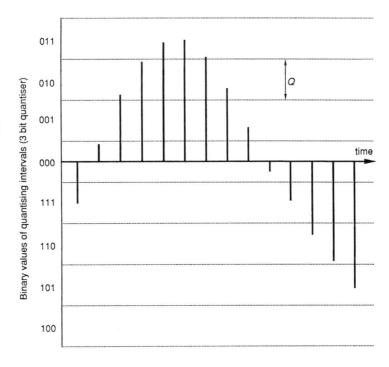

quantising steps is equal to 2^n where n is the number of bits in the binary words used to represent each sample. Consequently, a 4 bit quantiser offers only 2^4 (16) quantising levels, whereas a 16 bit quantiser offers a much larger 2^{16} or 65 536 levels.

Clearly an error may be involved in quantisation, since there are only a limited number of discrete levels available to represent the amplitude of the signal at any time. The error size will be a maximum of plus or minus half the amplitude of one quantising step, and a greater number of bits per sample will therefore result in a smaller error (see Figure 2.9), provided that the analog voltage range represented remains the same.

Figure 2.10 shows the binary number range covered by digital audio signals at different resolutions, using the usual two's complement representation. It will be seen that the maximum positive sample value of a 16 bit signal is &7FFF, whilst the maximum negative value is &8000. The sample value changes from all zeros (&0000) to all ones (&FFFF) as it crosses the zero volts point. The maximum digital signal level is normally termed 0 dBFS (FS = full scale). Signals rising above this level are normally hard-clipped, resulting in severe distortion, as shown in Figure 2.11.

Figure 2.9 In (a) a 3 bit
scale is used and only a
small number of quantising
intervals covers the analog
voltage range, making the
maximum quantising error
quite large. The second
sample in this picture will be
assigned the value 010, for
example, the corresponding
voltage of which is somewhat
higher than that of the
sample. During D/A
conversion the binary sample
values from (a) would be
turned into pulses with the
amplitudes shown in (b),
where many samples have
been forced to the same
level owing to quantising. In
(c) the 4 bit scale means that
a larger number of intervals
is used to cover the same
range and the quantising
error is reduced. (Expanded
positive range only shown for
clarity)

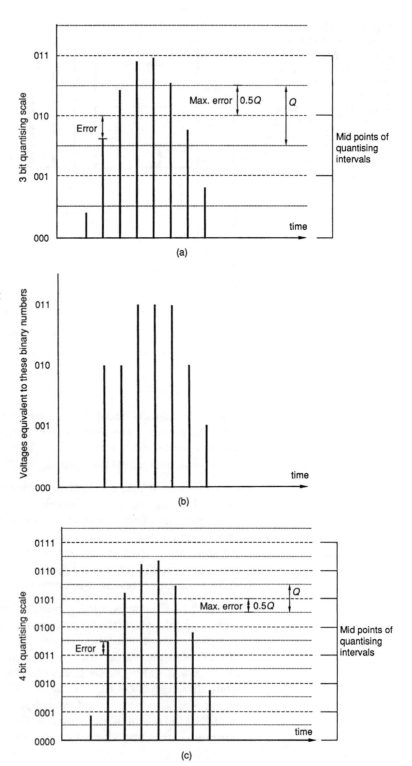

Figure 2.10 Binary number ranges (in hexadecimal) related to analog voltage ranges for different convertor resolutions, assuming two's complement representation of negative values. (a) 8 bit quantiser; (b) 16 bit quantiser; (c) 20 bit quantiser

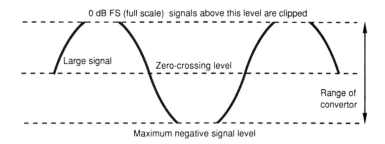

	(a)	(b)	(c)
Max. +ve signal voltage	7F	7FFF	7FFFF
		Positive values	
Zero volts	00	0000	00000
	FF	FFFF	FFFFF
		Negative values	
Max. −ve signal voltage	80	8000	80000

Figure 2.11 Signals exceeding peak level in a digital system are hard-clipped, since no more digits are available to represent the sample value

0 dB FS (full scale) signals above this level are clipped

Large signal   Zero-crossing level

Range of convertor

Maximum negative signal level

Figure 2.12 Quantising error depicted as an unwanted signal added to the original sample values. Here the error is highly correlated with the signal and will appear as distortion. (Courtesy of Allen Mornington West)

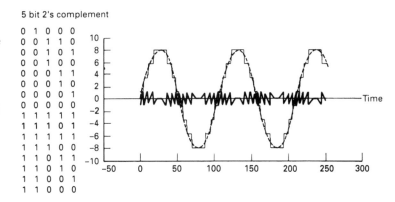

5 bit 2's complement

```
0 1 0 0 0
0 0 1 1 0
0 0 1 0 1
0 0 1 0 0
0 0 0 1 1
0 0 0 1 0
0 0 0 0 1
0 0 0 0 0
1 1 1 1 1
1 1 1 0 1
1 1 1 1 1
1 1 1 0 0
1 1 0 1 1
1 1 0 1 0
1 1 0 0 1
1 1 0 0 0
```

2.1.4 Audible effects of sample resolution

The quantising error may be considered as an unwanted signal added to the wanted signal, as shown in Figure 2.12. Unwanted signals tend to be classified either as distortion or noise, depending on their characteristics, and the nature of the quantising error signal depends very much upon the level and nature of the related audio signal. Here are a few examples, the illustrations for which have been prepared in the digital domain for clarity, using 16 bit sample resolution.

(a)

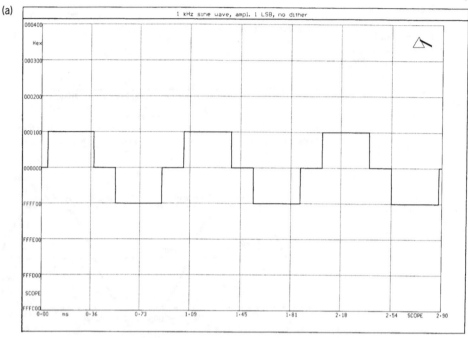

(b)

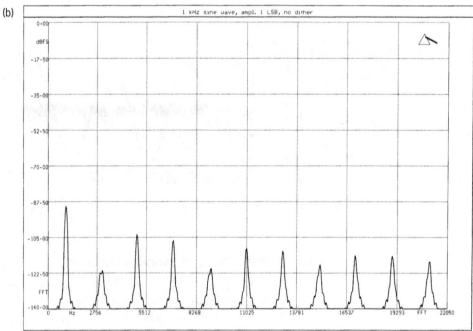

Figure 2.13 (a) A 1 kHz sine wave at very low level (amplitude ±1 LSB) just turns the least significant bit of the quantiser on and off. Analysed in the digital domain with sample values shown in hex on the vertical axis and time in ms on the horizontal axis. (b) Frequency spectrum of this quantised sine wave, showing distortion products

First consider a very low level sine wave signal, sampled then quantised, having a level only just sufficient to turn the least significant bit of the quantiser on and off at its peak (see Figure 2.13(a)). Such a signal would have a quantising error that was periodic, and strongly correlated with the signal, resulting in harmonic distortion. Figure 2.13(b) shows the frequency spectrum, analysed in the digital domain of such a signal, showing clearly the distortion products (predominantly odd harmonics) in addition to the original fundamental. Once the signal fell below the level at which it just turned on the LSB there would be no modulation. The audible result, therefore, of fading a signal down to silence would be that of an increasingly distorted signal suddenly disappearing. A higher level sine wave signal would cross more quantising intervals and result in more non-zero sample values. As the signal level rises the quantising error, still with a maximum value of ±0.5Q, becomes increasingly small as a proportion of the total signal level, and the error gradually loses its correlation with the signal.

Consider now a music signal of reasonably high level. Such a signal has widely varying amplitude and spectral characteristics and consequently the quantising error is likely to have a random nature. In other words it will be more noise-like than distortion-like, hence the term *quantising noise* which is often used to describe the audible effect of quantising error. An analysis of the power of the quantising error, assuming that it has a noise-like nature, shows that it has an r.m.s. amplitude of $Q/\sqrt{12}$, where Q is the voltage increment represented by one quantising interval. Consequently the signal-to-noise (S/N) ratio of an ideal n bit quantised signal can be shown to be:

$$6.02\ n\ + 1.76\ \text{dB}$$

This implies a theoretical S/N ratio which approximates to just over 6 dB per bit. So a 16 bit convertor might be expected to exhibit an S/N ratio of around 98 dB, and an 8 bit convertor around 50 dB. This assumes an undithered convertor, which is not the normal case, as described below. If a convertor is undithered there will only be quantising noise when a signal is present, but there will be no quiescent noise floor in the absence of a signal. Issues of dynamic range with relation to human hearing are discussed further in section 2.3.1.

2.1.5 Use of dither

The use of dither in A/D conversion, and in conversion between one sample resolution and another, is now widely accepted as

Figure 2.14 (a) Dither noise added to a sine wave signal prior to quantisation. (b) Post-quantisation the error signal is noise-like and has no clear correlation with the original signal (heavy line). (Courtesy of Allen Mornington West)

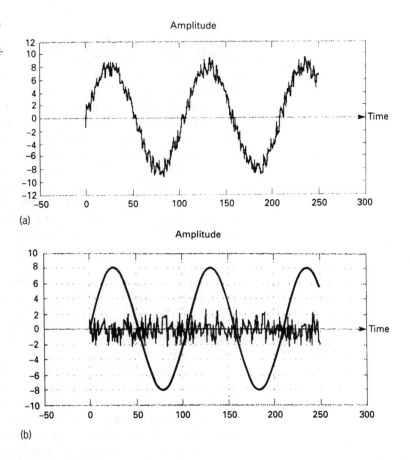

correct. It has the effect of linearising a normal convertor (in other words it effectively makes each quantising interval the same size), and turns quantising distortion into a random, noise-like signal at all times. This is desirable for a number of reasons. First because white noise at very low level is less subjectively annoying than distortion; second because it allows signals to be faded smoothly down to silence without the sudden disappearance noted above; and third because it often allows signals to be reconstructed even when their level is below the noise floor of the system. Undithered audio signals begin to sound 'grainy' and distorted as the signal level falls. Quiescent hiss will disappear if dither is switched off, making a system seem 'quieter', but it is normally considered that a small amount of continuous hiss is vastly preferable to low level distortion.

Dithering a convertor involves the addition of a very low level signal to the audio whose amplitude depends upon the type of dither employed (see below). The dither signal is usually noise, but may also be a waveform at half the sampling frequency, or a combination of the two. A signal which has not been correctly dithered during the A/D conversion process cannot thereafter be dithered with the same effect, since the signal will have been irrevocably distorted. How then does dither perform the seemingly remarkable task of removing quantising distortion?

It was stated above that the distortion was a result of the correlation between the signal and the quantising error, making the error periodic and subjectively annoying. Adding noise, which is a random signal, to the audio has the effect of randomising the quantising error and making it noise-like as well (shown in Figure 2.14 (a) and (b). If the noise has an amplitude similar in level to the LSB (in other words, one quantising step) then a signal lying exactly at the decision point between one quantising interval and the next may be quantised either upwards or downwards, depending on the instantaneous level of the dither noise added to it. Over time this random effect is averaged, leading to a noise-like quantising error and a fixed noise floor in the system.

 Figure 2.15(a) shows the same low level sine wave as in Figure 2.13, but this time with dither noise added. The quantised signal retains the cyclical pattern of the 1 kHz sine wave but is now modulated much more frequently between states, and a random element has been added. The frequency spectrum of this signal, Figure 2.15(b), shows a single sine wave component accompanied by a flat noise floor. Figure 2.15(c) and (d) show the waveform and spectrum of a dithered sine wave at a level which would be impossible to represent in an undithered 16 bit system. The LSB is in the zero state much more frequently, but an element of the original 1 kHz period can still be seen in its modulation pattern if studied carefully. The duty cycle of the LSB modulation (ratio between time on and time off) varies with the amplitude of the original signal. When this is passed through a D/A convertor and reconstruction filter the result is a pure sine wave signal plus noise, as can be seen from the spectrum analysis.

Dither is also used in digital processing devices such as mixers, but in such cases it is introduced in the digital domain as a random number sequence (the digital equivalent of white noise). In this context it is used to remove low-level distortion in signals whose gains have been altered, and to optimise the conversion from high resolution to lower resolution during post-production (see below).

2.1.6 Types of dither

Research, particularly by Vanderkooy and Lipshitz[2], has shown that certain types of dither signal are more suitable than others

(a)

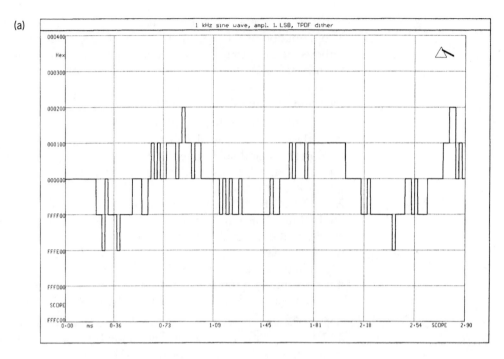

(b)

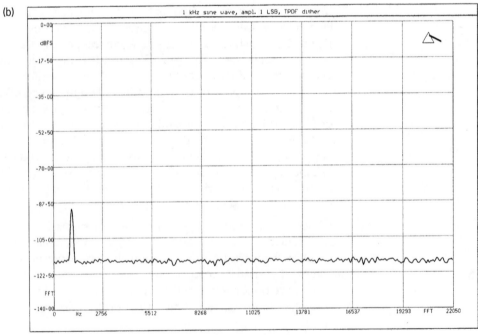

Figure 2.15 (a) 1 kHz sine wave, amplitude ±1 LSB, with dither added, analysed in the digital domain. (b) Spectrum of this dithered low level sine wave showing lack of distortion and flat noise floor. (c) 1 kHz sine wave at a level of −104 dBFS

(c)

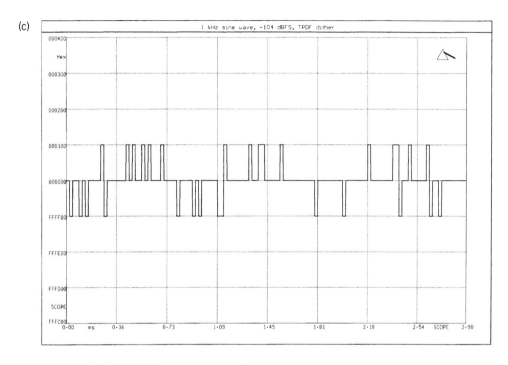

(d)

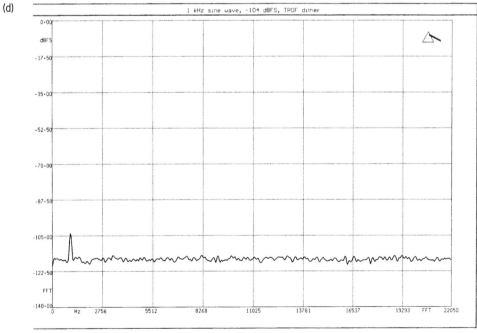

with dither, showing occasional modulation of LSB. (d) Spectrum of this signal showing that it is still possible to discern the original signal. An undithered 16 bit system would be incapable of representing a signal below about –97 dBFS

Figure 2.16 A probability distribution curve for dither shows the likelihood of the dither signal having a certain amplitude, averaged over a long time period

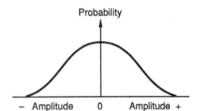

for high quality audio work. Dither noise is often characterised in terms of its probability distribution, which is a statistical method of showing the likelihood of the signal having a certain amplitude. A simple graph such as that shown in Figure 2.16 is used to indicate the shape of the distribution. The probability is the vertical axis and the amplitude in terms of quantising steps is the horizontal axis.

Logical probability distributions can be understood simply by thinking of the way in which dice fall when thrown (see Figure 2.17). A single die throw has a rectangular probability distribution function (RPDF), because there is an equal chance of the throw being between 1 and 6 (unless the die is weighted!). The total value of a *pair* of dice, on the other hand, has a roughly triangular probability distribution function (TPDF) with the peak grouped on values from 6 to 8, since there are more combinations that make these totals than there are combinations making 2 or 12. Going back to digital electronics, one could liken the dice to random number generators and see that RPDF dither could be created using a single random number generator, and that TPDF dither could be created by adding the outputs of two RPDF generators.

Figure 2.17 Probability distributions of dice throws. (a) A single die throw shows a rectangular PDF (b) A pair of thrown dice added together has a roughly triangular PDF (in fact it is stepped)

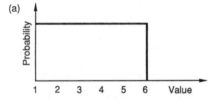

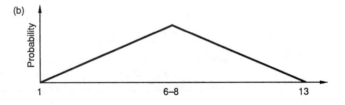

Figure 2.18 Most suitable digital dither signals for audio. (a) TPDF dither with a peak-to-peak amplitude of 2 Q. (b) RPDF dither with an amplitude of 1 Q.

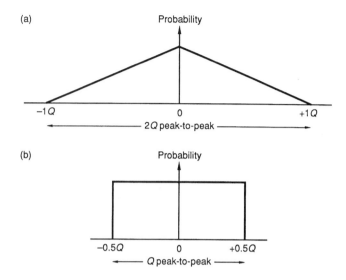

RPDF dither has equal likelihood that the amplitude of the noise will fall anywhere between zero and maximum, whereas TPDF dither has greater likelihood that the amplitude will be zero than that it will be maximum. Analog white noise has Gaussian probability, whose shape is slightly more unusual than either of the logically generated dithers. Lipshitz and Vanderkooy have demonstrated that whilst both RPDF, TPDF and Gaussian dither can have the effect of linearising conversion and removing distortion, RPDF dither tends to result in noise modulation at low signal levels. This leads them to propose that the most suitable dither noise is TPDF with a peak-to-peak amplitude of 2 Q (see Figure 2.18). If RPDF dither is used it should have a peak-to-peak amplitude of 1 Q.

Whilst it is easy to generate ideal logical PDF's in the digital domain, it is likely that the noise source present in many convertors will be analog, and therefore Gaussian in nature. With Gaussian noise, the optimum r.m.s. amplitude for the dither signal is 0.5 Q, at which level noise modulation is minimised but not altogether absent. Dither at this level has the effect of reducing the undithered dynamic range by about 6 dB, making the dithered dynamic range of an ideal 16 bit convertor around 92 dB.

2.1.7 Oversampling in A/D conversion

Oversampling involves sampling audio at a higher frequency than strictly necessary to satisfy the Nyquist criterion. Normally, though, this high rate is reduced to a normal rate in a subsequent digital filtering process, in order that no more storage space is

Figure 2.19 Block diagram
of oversampling A/D
conversion process

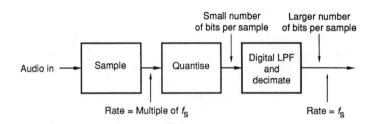

required than for conventionally sampled audio. It works by
trading off sample resolution against sampling rate, based on the
principle that the information carrying capacity of a channel is
related to the product of these two factors. Samples at a high rate
with low resolution can be converted into samples at a lower rate
with higher resolution, with no overall loss of information (sound
quality). Oversampling has now become so popular that it is the
norm in most high quality audio convertors.

Although oversampling A/D convertors often quote very high
sampling rates of up to 128 times the basic rates of 44.1 or 48
kHz, the actual rate at the digital output of the convertor is no
more than the basic rate. Samples acquired at the high rate are
quantised to only a few bits resolution and then digitally filtered
to reduce the sampling rate, as shown in Figure 2.19. The digital
low-pass filter limits the bandwidth of the signal to half the basic
sampling frequency in order to avoid aliasing, and this is
coupled with 'decimation'. Decimation reduces the sampling
rate by dropping samples from the oversampled stream. A result
of the low-pass filtering operation is to increase the word length
of the samples very considerably. This is not simply an arbitrary
extension of the wordlength, but an accurate calculation of the
correct value of each sample, based on the values of surround-
ing samples (see the section on digital signal processing, below).
Although oversampling convertors quantise samples initially at
a low resolution, the output of the decimator consists of samples
at the nominal Nyquist sampling rate with more bits of resolution. The sample resolution can then be shortened as necessary
(see the section on *requantising* , below) to produce the desired
word length.

Oversampling brings with it a number of benefits and is the key
to improved sound quality at both the A/D and D/A ends of a
system. Because the initial sampling rate is well above the audio
range (often tens or hundreds of times the nominal rate) the
spectral repetitions resulting from PAM are a long way from the
upper end of the audio band (see Figure 2.20). The analog anti-
aliasing filter used in conventional convertors is replaced by the

Figure 2.20 (a) Oversampling in A/D conversion initially creates spectral repetitions that lie a long way from the top of the audio baseband. The dotted line shows the theoretical extension of the baseband and the potential for aliasing, but the audio signal only occupies the bottom part of this band. (b) Decimation and digital low-pass filtering limits the baseband to half the sampling frequency, thereby eliminating any aliasing effects, and creates a conventional collection of spectral repetitions at multiples of the sampling frequency

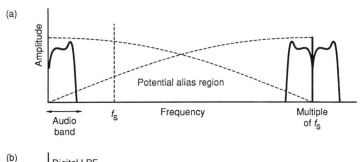

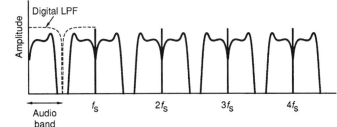

digital decimation filter, and if oversampling is also used in D/A conversion the analog reconstruction filter can have a shallower roll-off. This can have the effect of improving phase linearity within the audio band, which is known to improve audio quality. Oversampling also makes it possible to introduce so-called 'noise shaping' into the conversion process, which allows quantising noise to be shifted out of the most audible parts of the spectrum.

2.1.8 Noise shaping in A/D conversion

Noise shaping is a means by which noise within the most audible parts of the audio frequency range is reduced at the expense of increased noise at other frequencies, using a process which 'shapes' the spectral energy of the quantising noise. It is possible because of the high sampling rates used in oversampling convertors, since a high sampling rate extends the frequency range over which quantising noise is spread, putting much of it outside the audio band.

Quantising noise energy extends over the whole baseband, up to the Nyquist frequency. Oversampling spreads the quantising noise energy over a wider spectrum, since in oversampled convertors the Nyquist frequency is well above the upper limit of the audio band. This has the effect of reducing the in-band noise by around 3 dB per octave of oversampling (in other words, a system oversampling at twice the Nyquist rate would see the noise power within the audio band reduced by 3 dB).

45

Figure 2.21 Block diagram of a noise shaping delta-sigma A/D convertor

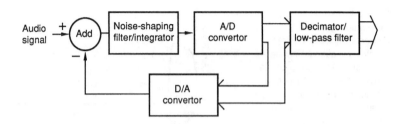

In oversampled noise-shaping A/D conversion an integrator (low-pass filter) is introduced before the quantiser, and a D/A convertor is incorporated into a negative feedback loop, as shown in Figure 2.21. This is the so-called 'sigma-delta convertor'. Without going too deeply into the principles of such convertors, the result is that the quantising noise (introduced after the integrator) is given a rising frequency response at the input to the decimator, whilst the input signal is passed with a flat response. There are clear parallels between such a circuit and analog negative-feedback circuits.

Without noise shaping, the energy spectrum of quantising noise is flat up to the Nyquist frequency, but with first-order noise shaping this energy spectrum is made non-flat, as shown in Figure 2.22. With second-order noise shaping the in-band reduction in

Figure 2.22 Frequency spectra of quantising noise. In a non-oversampled convertor, as shown in (a), the quantising noise is constrained to lie within the audio band. In an oversampling convertor, as shown in (b), the quantising noise power is spread over a much wider range, thus reducing its energy in the audio band. (c) With noise shaping the noise power within the audio band is reduced still further, at the expense of increased noise outside that band

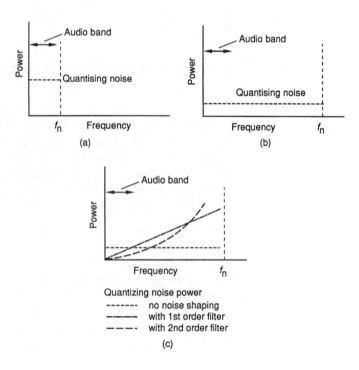

noise is even greater, such that the in-band noise is well below that achieved without noise shaping.

2.2 D/A conversion

2.2.1 A basic D/A convertor

The basic D/A conversion process is shown in Figure 2.23. Audio sample words are converted back into a staircase-like chain of voltage levels corresponding to the sample values. This is achieved in simple convertors by using the states of bits to turn current sources on or off, making up the required pulse amplitude by the combination of outputs of each of these sources. This staircase is then 'resampled' to reduce the width of the pulses before they are passed through a low-pass reconstruction filter, whose cut-off frequency is half the sampling frequency. The effect of the reconstruction filter is to join up the sample points to make a smooth waveform. Resampling is necessary because otherwise the averaging effect of the filter would result in a reduction in the amplitude of high-frequency audio signals (the so-called 'aperture effect'). Aperture effect may be reduced by limiting the width of the sample pulses to perhaps one-eighth of the sample period. Equalisation may be required to correct for aperture effect.

2.2.2 Oversampling in D/A conversion

Oversampling may be used in D/A conversion, as well as in A/D conversion. In the D/A case additional samples must be created in between the Nyquist rate samples, in order that conversion can be performed at a higher sampling rate. These are produced by sample rate conversion of the PCM data. (Sample rate conversion is introduced in a later section.) These samples are then converted back to analog at the higher rate, again avoiding the need for steep analog filters. Noise shaping may also be introduced at the D/A stage, depending on the

Figure 2.23 Processes involved in D/A conversion (positive sample values only shown)

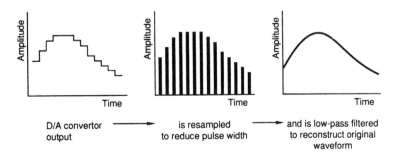

D/A convertor output ⟶ is resampled to reduce pulse width ⟶ and is low-pass filtered to reconstruct original waveform

design of the convertor, to reduce the subjective level of the noise.

A number of advanced D/A convertor designs exist which involve oversampling at a high rate, creating samples with only a few bits of resolution. The extreme version of this approach involves very high rate conversion of single bit samples (so-called 'bit stream conversion'), with noise shaping to optimise the noise spectrum of the signal. The theory of these convertors is outside the scope of this book.

2.3 Sound quality versus sample rates and resolutions

The question often arises as to what sample rate and resolution is necessary for a certain quality of audio. What are the effects of selecting certain values? Are there standards? This section aims to provide some guidelines in this area, with reference to the capabilities of human hearing which must be considered the ultimate arbiter in this matter.

2.3.1 Psychoacoustic limitations

It is possible with digital audio to approach the limits of human hearing in terms of sound quality. It is also true, though, that badly engineered digital audio can sound poor, and that the term 'digital' does not automatically imply high quality. The choice of sampling parameters and noise shaping methods affects the frequency response, distortion and perceived dynamic range of digital audio.

The human ear's capabilities should be regarded as the standard against which the quality of digital systems are measured, since it could be argued that the only distortions and noises which matter are those that can be heard. It might be considered wise to design a convertor whose noise floor was tailored to the low level sensitivity of the ear, for example. Figure 2.24 shows a typical low level hearing sensitivity curve, indicating the sound pressure level (SPL) required for a sound just to be audible. It will be seen that the ear is most sensitive in the middle frequency range, around 4 kHz, and that the response tails off towards the low and high frequency ends of the spectrum. This curve is often called the 'minimum audible field (MAF)' or 'threshold of hearing'. It has an SPL of 0 dB (ref. 20 µPa) at 1 kHz. It is worth remembering, though, that the thresholds of hearing of the human ear are not absolute but probabilistic. In other words, when trying to determine what can and cannot be

Figure 2.24 Hearing threshold curve

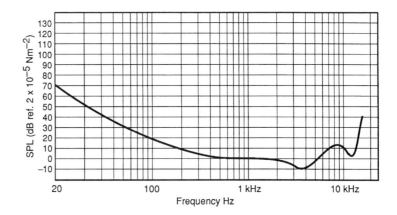

perceived one is dealing with statistical *likelihood* of perception. This is important for any research which attempts to establish criteria for audibility, since there are certain sounds which, although as much as 10 dB below the accepted thresholds, have a statistical likelihood of perception which may approach certainty in some cases.

Dynamic range could be said to be equal to the range between the MAF and the loudest sound tolerable. The loudest sound tolerable depends very much on the person, but the threshold of 'pain' is usually said to occur between 130 and 140 dB SPL. The absolute maximum dynamic range of human hearing is therefore around 140 dB at 1 kHz, but quite a lot less than that at low and high frequencies. Whether or not it is desirable to be able to record and reproduce such a wide dynamic range is debatable.

Work carried out by Louis Fielder and Elizabeth Cohen[3,4] attempted to establish the dynamic range requirements for high quality digital audio systems by investigating the extremes of sound pressure available from acoustic sources and comparing these with the perceivable noise floors in real acoustic environments. Using psychoacoustic theory, Fielder was able to show what was likely to be heard at different frequencies in terms of noise and distortion, and where the limiting elements might be in a typical recording chain. Having defined dynamic range as 'the ratio between the r.m.s. maximum undistorted sine wave level producing peak levels equal to a particular peak level and the r.m.s. level of 20 kHz band-limited white noise that has the same apparent loudness as a particular audio chain's equipment noise in the absence of a signal', he proceeded to show that the just audible level of a 20 kHz bandwidth noise signal was about 4 dB SPL, and that a number of musical performances reached levels of between 120 and 129 dB SPL in favoured listening positions. From this he determined a dynamic range requirement of 122 dB for natural reproduction. Taking into account microphone performance and the limitations of consumer loudspeakers, this requirement dropped to 115 dB for consumer systems.

2.3.2 Sampling rate

The choice of sampling rate determines the maximum audio bandwidth available. There is a strong argument for choosing a sampling rate no higher than is strictly necessary, in other words not much higher than twice the highest audio frequency to be represented. This often starts arguments over what is the highest useful audio frequency, and it is an area over which heated debates have raged. Conventional wisdom has it that the audio frequency band extends up to 20 kHz, and this implies the need for a sampling frequency of just over 40 kHz for high quality audio work. There are in fact two standard sampling frequencies between 40 and 50 kHz: the Compact Disc rate of 44.1 kHz, and the so-called 'professional' rate of 48 kHz. These are both allowed in the AES5 standard of 1984, which sets down preferred sampling frequencies for digital audio equipment. Table 2.1 is an attempt to summarise the variety of sampling frequencies in existence and their applications. A wide range of rates is used in desktop PCs and sound cards, and only the most popular are shown here.

The 48 kHz rate was originally included because it left a certain amount of leeway for downward varispeed in tape recorders. When many digital recorders are varispeeded, their sampling rate changes proportionately and the result is a shifting of the first spectral repetition of the audio baseband. If the sampling rate is reduced too far aliased components may become audible. Most professional digital tape recorders allowed for only around ±12.5 per cent of varispeed for this reason. It is possible now, though, to avoid such problems using digital low pass filters whose cut-off frequency varies with the sampling frequency.

The 44.1 kHz frequency had been established earlier on for the consumer Compact Disc (CD), and is very widely used in the industry. In fact in many ways it has become the sampling rate of choice for most professional recordings. It allows for full use of the 20 kHz audio band, and oversampling convertors allow for the use of shallow analog anti-aliasing filters which avoid phase problems at high audio frequencies. It also generates 10 per cent less data per second than the 48 kHz rate, making it economical from a storage point of view.

Table 2.1 Commonly encountered sampling frequencies

Frequency (kHz)	Application
8	Used in telephony. Poor audio quality. CCITT G711 standard. IMA RP rate*
~11.025	One quarter of the CD sampling rate, used in the sound hardware of some desktop computers, particularly the Apple Macintosh, for low quality applications. IMA RP rate*
16	Used in some telephony applications. G722 data reduction
18.9	CD-ROM/XA and CD-I standard for low–moderate quality audio using ADPCM to extend playing time
~22.05	Half the CD rate is 22.05 kHz. The original Apple Macintosh audio sampling rate was 22254.5454..... IMA RP rate*
32	Used in some broadcast systems, e.g. NICAM 3, NICAM 728, DAT long play mode
37.8	CD-ROM/XA and CD-I sampling rate for intermediate quality audio using ADPCM
44.056	A slight modification of the 44.1 rate used in some older equipment to align digital audio with the NTSC television frame rate of 29.97 frames per second. Occasionally still encountered in the USA
44.1	CD sampling rate. Used widely for professional audio recording in many formats. Some professionally modified DAT machines will operate at this rate from analog inputs. IMA RP rate*
47.952	Occasionally encountered when 48 kHz equipment is used in NTSC video operations. To be avoided
48	'Professional' rate, as specified in AES5-1984, and encountered mainly in digital video recorder sound tracks. Many DAT machines will only sample at this rate through analog inputs
88.2 and 96	Twice the 44.1 and 48 k standard rates. Found in some audiophile equipment, such as certain non-standard DAT machines. There is currently a certain amount of pressure to standardise the 96 kHz rate for very high quality applications

*IMA RP rates were selected in the International Multimedia Association Recommended Practice for Enhancing Digital Audio Compatibility in Multimedia Systems, Rev. 3.00, October 1992, for sound file transfer between workstations

A rate of 32 kHz is used in some broadcasting applications, such as NICAM 728 stereo TV transmissions, and in some radio distribution systems. Television and FM radio sound bandwidth has been limited to 15 kHz for many years, and a considerable economy of transmission bandwidth is achieved by the use of this lower sampling rate. The majority of important audio information lies below 15 kHz in any case, and little is lost by removing the top 5 kHz of the audio band. Some professional audio applications offer this rate as an option, but it is not common. It is used for the long play mode of some DAT machines, for example.

Arguments have been presented for the standardisation of higher sampling rates such as 88.2 and 96 kHz, quoting evidence from sources claiming that information above 20 kHz is important for sound quality. One Japanese professor[5] has claimed that frequencies above 20 kHz stimulate the production of so-called alpha waves in the brain which correspond with a state of satisfaction and relaxation. It is certainly true that the ear's frequency response does not cut off completely at 20 kHz, but there is very little properly supported evidence that listeners can repeatably distinguish between signals containing higher frequencies and those which do not. Doubling the sampling frequency leads to a doubling in the overall data rate of a digital audio system, and a consequent halving in storage time. It follows that these higher sampling rates should be used only after careful consideration of the merits. Remember that there is currently no standard consumer replay medium that supports these rates.

Low sampling frequencies such as those below 30 kHz are sometimes encountered in PC workstations for lower quality sound applications such as the storage of speech samples, the generation of internal sound effects and so forth. They are the result of the limited clock and processing speeds available in earlier computers. Multimedia applications may need to support these rates because such applications often involve the incorporation of sounds of different qualities.

2.3.3 Quantising resolution

The number of bits per sample dictates the signal-to-noise ratio or dynamic range of a digital audio system. For the time being only linear PCM systems will be considered, because the situation is different when considering systems which use non-uniform quantisation or data reduction (see Chapter 3). Table 2.2 attempts to summarise the applications for different sample resolutions.

For many years now 16 bit linear PCM has been considered the norm for high quality audio applications. This is the CD standard and is capable of offering good dynamic range of over 90 dB. For most purposes this is adequate, but it fails to reach Fielder's ideal (quoted above) of 122 dB for subjectively noise-free reproduction in professional systems. To achieve such a dynamic range requires a convertor resolution of around 21 bits, which is nearly achievable with today's convertor technology, depending on how the specification is interpreted. Some designs employ two convertors with a gain offset, using digital signal processing to combine the outputs of the two in the range where they overlap, achieving a significant increase in the perceived

Table 2.2 Linear quantising resolution

Bits per sample	Approx dyn. range with dither	Application
8	44 dB	Low–moderate quality for older PC internal sound generation. Some multimedia applications. Usually in the form of unsigned binary numbers
12	68 dB	Older Akai samplers, e.g. S900
14	80 dB	Original EIAJ format PCM adaptors, such as Sony PCM-100
16	92 dB	CD standard. DAT standard. Most widely used high quality resolution for consumer media and many professional recorders. Many multimedia PCs. Two's complement (signed) binary numbers
20	116 dB	High quality professional audio recording and mastering applications. Good convertors available
24	140 dB	Maximum resolution of most new professional recording systems, also of AES/EBU digital interface. Dynamic range would exceed psychoacoustic requirements. Hard to convert at this resolution

dynamic range. Others use two convertors in parallel with independent dither, summing their outputs so that the signal rises by 6 dB but the noise by only 3 dB.

It is often the case that for professional recording purposes one needs a certain amount of 'headroom' – in other words some unused dynamic range above the normal peak recording level which can be used in unforeseen circumstances such as when a signal overshoots its expected level. This can be particularly necessary in live recording situations where one is never quite sure what is going to happen with recording levels. This is another reason why many professionals feel that a resolution of greater than 16 bits is desirable for original recording. 20 bit recording formats are becoming increasingly popular for this reason, with mastering engineers then optimising the finished recording for 16 bit media (such as CD) using noise-shaped requantising processes (see section 2.5). There is even the beginning of interest in 24 bit recording, but there is currently no conversion technology capable of fully exploiting this dynamic range.

At the lower quality end, some PC sound cards and internal sound generators operate at resolutions as low as 4 bits. 8 bit

Table 2.3 IMA recommended sampling rates and quantisations for basic audio interchange on computers

Sampling rate	Mono/stereo	Quantisation	Notes
8 kHz	Mono	8 bit μ-law PCM	CCITT G.711 standard
	Mono	8 bit A-law PCM	CCITT G.711 standard
	Mono	4 bit ADPCM	Intel/DVI algorithm
11.025 kHz	Mono/stereo	8 bit linear PCM	Apple Macintosh and MPC
	Mono/stereo	4 bit ADPCM	Intel/DVI algorithm
22.05 kHz	Mono/stereo	8 bit linear PCM	Apple Macintosh and MPC
	Mono/stereo	4 bit ADPCM	Intel/DVI algorithm
44.1 kHz	Mono/stereo	16 bit linear PCM	LSByte first
	Mono/stereo	16 bit linear PCM	MSByte first
	Mono/stereo	4 bit ADPCM	DVI algorithm

MPC is the Microsoft Multimedia PC standard. CCITT standards are issued by the ITU Telecommunications Bureau. The Intel DVI algorithm is described in more detail in Chapter 3.

resolution is quite common in desktop computers, and this proves adequate for moderate quality sound through the PC's internal loudspeakers. It gives a dynamic range of nearly 50 dB undithered. There are a number of non-linear quantisation schemes for PCs, such as A-law and μ-law, as well as various data compression schemes, which give an improved dynamic range from a small number of bits but with audible side effects. The International Multimedia Association has standardised a set of recommended sampling rates and data types for the purpose of simplifying interchange, as shown in Table 2.3. Modern multimedia PCs and sound cards generally offer 16 bit resolution as standard. Some early MIDI samplers operated at 8 bit resolution, and some more recent models at 12 bit, but it is now common for MIDI samplers to offer full 16 bit resolution.

2.4 Timing jitter and its effect on convertors

The issue of timing stability of digital audio signals has become more prominent recently as users and designers have come to realise that this is just as important for sound quality as the number of bits available. It is a large subject, and it is not proposed to examine it in detail here, but a summary will be given.

Timing jitter is the short-term variation of the timing position of audio samples. Ideally they should be exactly evenly spaced. If they are not then various detrimental effects are noticed in the sound quality, including additional noise and distortion. This depends on the manner in which the sample instants are time-shifted. Jitter may be considered as very similar to quantising

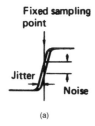

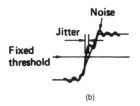

Figure 2.25 (a) Short-term timing errors are termed jitter. (b) Noise on the signal can also appear as timing jitter when compared with a fixed threshold

error, but on the time axis instead of the amplitude axis (see Figure 2.25), and it has similar effects. It is caused by a variety of factors including poor quality clock signals, electrical noise, interference, and often by carrying audio over digital interfaces. It is important to note that jitter is not necessarily a problem. It only affects sound quality if it affects the clock used in conversion to and from the analog domain. In many cases it can be removed by a process of reclocking using a suitably stable phase-locked loop (an electrical circuit which acts to some extent like a heavy flywheel, and is capable of ironing out short-term irregularities in clock signals). Jitter on the A/D convertor clock cannot subsequently be removed and creates permanent distortion. Jitter introduced at later stages in the digital domain may be reduced or removed.

The important features of jitter are its peak amplitude and its rate, since the effect on sound quality is dependent on both of these factors taken together. Shelton, by calculating the r.m.s. signal-to-noise ratio resulting from random jitter, shows that timing irregularities as low as 5 ns may be significant for 16 bit digital audio systems over a range of signal frequencies, and that the criteria are even more stringent at higher resolutions and at high frequencies. The effects are summarized in Figure 2.26.

When jitter is periodic rather than random, it results in the equivalent of 'flutter', and the effect when applied to the sample clock in the conversion of a sinusoidal audio signal is to produce sidebands on either side of the original audio signal due to phase modulation, whose spacing is equal to the jitter frequency. Julian Dunn[6] has shown that the level of the jitter sideband (R_j) with relation to the signal is given by:

Figure 2.26 Effects of sample clock jitter on signal-to-noise ratio at different frequencies, compared with theoretical noise floors of systems with different resolutions (after W.T. Shelton, with permission)

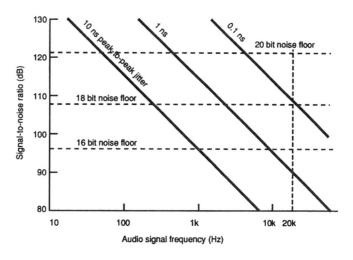

55

$$R_j \text{ (dB)} = 20 \log (J\omega_i/4)$$

where J is the peak-to-peak amplitude of the jitter and ω_i is the audio signal frequency. Using this formula he shows that for sinusoidal jitter with an amplitude of 500 ps, a maximum level 20 kHz audio signal will produce sidebands at –96.1 dB relative to the amplitude of the tone.

What is important, though, is the *audibility* of jitter-induced products, and Dunn[6,7] has attempted to calculate this based on an analysis of the resulting spectrum using psychoacoustic theory, assuming that the audio signal is replayed at a high listening level (120 dB SPL). As shown in Figure 2.27, which plots jitter amplitude against jitter frequency (not audio frequency) for just-audible modulation noise on a worst-case audio signal, the jitter amplitude may in fact be very high (>1 µs) at low jitter frequencies (up to around 250 Hz) since the sidebands will be masked at all audio frequencies, but the amount allowed falls sharply above this jitter frequency, although it may still be up to ±10 ns at jitter frequencies up to 400 Hz.

The benefits of using oversampling convertors are not entirely clear. Story[8] suggests that they reject sample clock jitter to a greater extent than non-oversampling convertors, because they reduce the bandwidth of the jitter by the oversampling factor, thus making them much more suitable than conventional convertors. Dunn, though, suggests that high frequency jitter (well above the audio band) in a delta-sigma DAC may modulate the shaped ultrasonic modulator noise, creating products within critical parts of the audio spectrum. A comprehensive study by Chris Dunn and Malcolm Hawksford[9] attempts

Figure 2.27 Sample clock jitter amplitude at different frequencies required for just audible modulation noise on a worst-case audio signal (after J. Dunn, with permission)

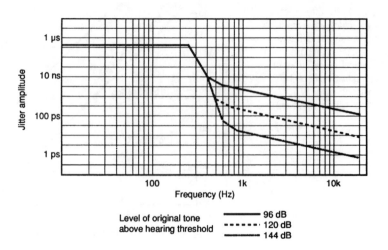

to survey the effects of digital-interface-induced jitter on different types of DAC, and this paper warrants close study by those whose business it is to design high quality DACs for audio applications.

2.5 Changing the resolution of an audio signal (requantisation)

There may be points in an audio production when the need arises to change the resolution of a signal. The most common example of this in high quality audio is when mastering CDs from, say, 20 bit recordings, since the CD has only 16 bit resolution, but it also occurs within signal processors of all types because sample wordlengths may vary at different stages. In multimedia applications it is common to need to convert sound files from 16 to 8 bit resolution. It is important that this operation is performed correctly, because incorrect requantisation results in unpleasant distortion.

If the length of audio samples needs to be reduced then the worst possible solution is simply to remove unwanted LSBs. Taking the example of a 20 bit signal being reduced to 16 bits, one should not simply remove the 4 LSBs and expect everything to be alright. By removing the LSBs one would be creating a similar effect to not using dither in A/D conversion – in other words one would introduce low level distortion components. Low level signals would sound grainy and would not fade smoothly into noise. Figure 2.28 shows a 1 kHz signal which originally began life at 20 bit resolution, but has been truncated to 16 bits. The harmonic distortion is clearly visible.

The correct approach is to redither the signal for the target resolution by adding dither noise in the digital domain (see section 2.1.5). This digital dither should be at an appropriate level for the new resolution, and the LSB of the new sample should then be rounded up or down depending on the total value of the LSBs to be discarded, as shown in Figure 2.29. It is worrying to note how many low cost digital audio applications fail to perform this operation satisfactorily (particularly in 16 to 8 bit requantisation), leading to complaints about sound quality. Many professional quality digital audio workstations allow for audio to be stored and output at a variety of resolutions, and may make dither user selectable. They also allow the level of the audio signal to be changed in order that maximum use may be made of the available bits. It is normally important, for example, when mastering a CD from a 20 bit recording, to ensure that the highest level signal on the original recording is adjusted during

(a)

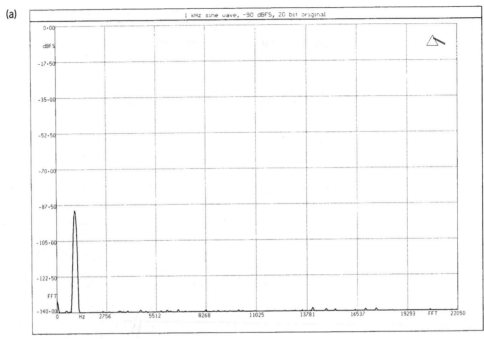

(b)

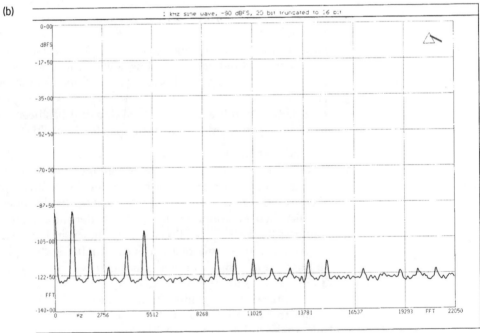

Figure 2.28 Truncation of audio samples results in distortion. (a) shows the spectrum of a 1 kHz signal generated and analysed at 20 bit resolution. In (b) the signal has been truncated to 16 bit resolution and the distortion products are clearly noticeable

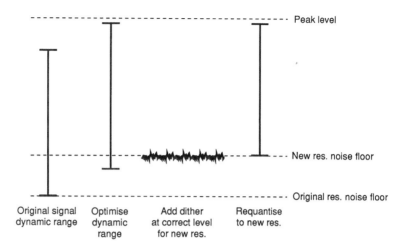

Figure 2.29 The correct order of events when requantising an audio signal at a lower resolution is shown here

mastering so that it peaks as close to the maximum level as possible *before* requantising and redithering at 16 bit resolution. In this way as much as possible of the original low level information is preserved and quantising noise is minimised. This applies in any requantising operation, not just CD mastering. A number of applications are available which automatically scale the audio signal so that its level is optimised in this way, allowing the user to set a peak signal value up to which the highest level samples will be scaled. Since some overload detectors on digital meters and CD mastering systems look for repeated samples at maximum level to detect clipping, it is perhaps wise to set peak levels so that they lie just below full modulation. This will ensure that master tapes are not rejected for a suspected recording fault by duplication plants, and subsequent users do not complain of 'over' levels.

2.6 Dynamic range enhancement

It is possible to maximise the subjective dynamic range of digital audio signals during the process of requantisation described above. This is particularly useful when mastering high resolution recordings for CD because the reduction to 16 bit wordlengths would normally result in increased quantising noise. It is in fact possible to retain most of the dynamic range of a higher resolution recording, even though it is being transferred to a 16 bit medium, and this remarkable feat is achieved by a noise shaping process similar to that described earlier.

During requantisation digital filtering is employed to shape the spectrum of the quantising noise so that as much of it as possible is shifted into the least audible parts of the spectrum. This

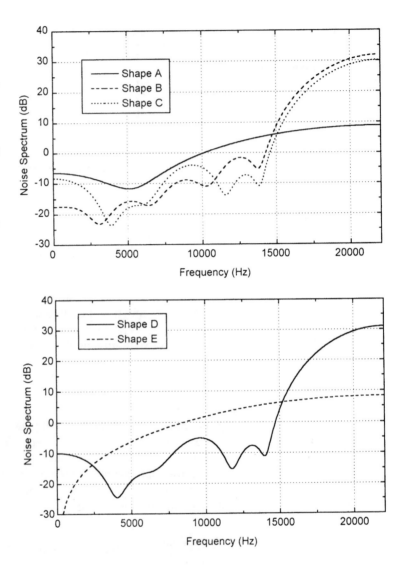

Figure 2.30 Examples of noise weighting curves used in the Meridian 518 mastering processor. Note linear frequency scale. Shape A: flat dither, 2nd order shaper. Shape B: flat dither, 9th order shaper (MAP). Shape C: flat dither, 9th order shaper (MAF). Shape D: high pass dither, 9th order shaper (MAF). Shape E: High pass dither. MAP = minimum audible pressure, MAF = minimum audible field. (Courtesy of J. R. Stuart and R. J. Wilson, Meridian Audio)

usually involves moving the noise away from the 4 kHz region where the ear is most sensitive and increasing it at the HF end of the spectrum. The result is often quite high levels of noise at HF, but still lying below the audibility threshold. In this way CDs can be made to sound almost as if they had the dynamic range of 20 bit recordings.[10] Some typical weighting curves used in a commercial mastering processor from Meridian are shown in Figure 2.30, although many other shapes are in use.

This is the principle employed in mastering systems such as Sony's Super Bit Mapping (SBM) and Deutsche Grammophon's Authentic Bit Imaging (ABI). Interestingly, some approaches allow the mastering engineer to choose from a number of 'shapes' of noise until he finds one which is subjectively the most pleasing for the type of music concerned, whereas others stick to one theoretically derived 'correct' shape.

2.7 Error correction

Since this book is concerned with digital audio for workstations the topic of error correction will only be touched upon briefly. Although dedicated audio recording formats need specially designed systems to protect against the effects of data errors, systems which use computer mass storage media do not. The reason for this is that mass storage media are formatted in such a way as to make them essentially error free. When, for example, a computer disk drive is formatted at a low level, the formatting application attempts to write data to each location and read it back. If the location proves to be damaged or gives erroneous replay it is noted as a 'bad block', after which it is never used for data storage. In addition, disk and tape drives look after their own error detection and correction by a number of means which are normally transparent to the digital audio system. If a data error is detected when reading data then the block of data is normally re-read a few times to see if the data can be retrieved. The only effect of this is to slightly slow down transfer.

This differs greatly from the situation with dedicated audio formats such as DAT. In dedicated audio formats there are many levels of error protection, some of which allow errors to be completely corrected (no effect on sound quality), and others which allow the audible effects of more serious errors to be minimised. The process known as interpolation, for example, allows missing samples to be 'guessed' by estimating the level of the sample based on those around it (see Figure 2.31). Computer systems, on the other hand, cannot allow this type of error correction because it is assumed that data is either correct or it is useless. When reading a financial spreadsheet, for example, it would not be acceptable for an erroneous figure to be guessed by looking at those on either side!

The result is that computer mass storage media are treated as raw, error-free data storage capacity, without the need to add an overhead for error correction data once formatted. This does not mean that such media are infallible and will never give errors, because they do fail occasionally, but that digital audio workstations do not normally use any additional procedures on top of those already in place. The downside of this is that if an unavoidable error does arise in the replay of a sound file from

Figure 2.31 Interpolation is a means of hiding the audible effects of missing samples, as shown here

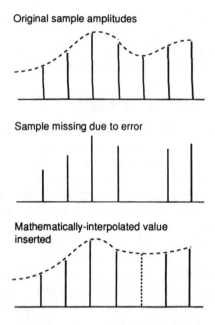

Original sample amplitudes

Sample missing due to error

Mathematically-interpolated value inserted

a digital workstation then it often results in a total inability to play that file. The file is assumed to be corrupted and the computer will not read it. The user does not have the option of being able to decide whether the error is audible, he must instead resort to one of the various computer file 'rescue packages' which attempt to rebuild the corrupted information using various proprietary techniques.

2.8 Introduction to digital audio signal processing

Just as processing operations like equalisation, fading and compression can be performed in the analog domain, so they can in the digital domain. Indeed it is often possible to achieve certain operations in the digital domain with fewer side effects such as phase distortion. It is possible to perform operations in the digital domain which are either very difficult or impossible in the analog domain. High quality, authentic-sounding artificial reverberation is one such example, in which the reflection characteristics of different halls and rooms can be accurately simulated. Digital signal processing (DSP) involves the high speed manipulation of the binary data representing audio samples. It may involve changing the values and timing order of samples, and it may involve the combining of two or more streams of audio data. DSP can affect the sound quality of digital audio in that it can add noise or distortion, although one must assume that the aim of good design is to minimise any such degradation in quality.

In the sections which follow an introduction will be given to some of the main applications of DSP in audio workstations without delving into the mathematical principles involved. In some cases the description is an over-simplification of the process, but the aim has been to illustrate concepts not to tackle the detailed design considerations involved.

2.8.1 Gain changing (level control)

It is relatively easy to change the level of an audio signal in the digital domain. It is easiest to shift its gain by 6 dB since this involves shifting the whole sample word either one step to the left or right (see Figure 2.32). Effectively the original value has been multiplied or divided by a factor of two. More precise gain control is obtained by multiplying the audio sample value by some other factor representing the increase or decrease in gain. The number of bits in the multiplication factor determines the accuracy of gain adjustment. The result of multiplying two binary numbers together is to create a new sample word which may have many more bits than the original, and it is common to find that digital mixers have internal structures capable of handling 32 bit words, even though their inputs and outputs may handle only 20. Because of this, redithering is usually employed in mixers at points where the sample resolution has to be shortened, such as at any digital outputs or conversion stages, in order to preserve sound quality as described above.

The values used for multiplication in a digital gain control may be derived from any user control such as a fader, rotary knob or on-screen representation, or they may be derived from stored values in an automation system. A simple 'old-fashioned' way of deriving a digital value from an 'analog' fader is to connect the fader to a fixed voltage supply and connect the fader wiper to an A/D convertor, although it is quite common now to find controls capable of providing a direct binary output relating to

Figure 2.32 The gain of a sample may be changed by 6 dB simply by shifting all the bits one step to the left or right

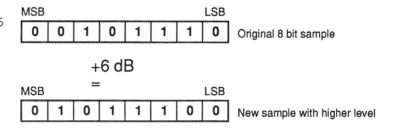

their position. The 'law' of the fader (the way in which its gain is related to its physical position) can be determined by creating a suitable look-up table of values in memory which are then used as multiplication factors corresponding to each physical fader position.

2.8.2 Crossfading

Crossfading is employed widely in digital audio workstations at points where one section of sound is to be joined to another (edit points). It avoids the abrupt change of waveform which might otherwise result in an audible click, and allows one sound to take over smoothly from the other. (This is discussed further in Chapter 4.)

The process is illustrated conceptually in Figure 2.33. It involves two signals each undergoing an automated fade (binary multiplication), one downwards and the other upwards, followed by an addition of the two signals. By controlling the rates and coefficients involved in the fades one can create different styles of crossfade for different purposes.

2.8.3 Mixing

Mixing is the summation of independent data streams representing the different audio channels. Time coincident samples from each input channel are summed to produce a single output channel sample. Clearly it is possible to have many mix 'buses' by having a number of separate summing operations for different output channels. The result of summing a lot of signals may be to increase the overall level considerably, and the architecture of the mixer must allow enough headroom for this possibility. In the same way as an analog mixer, the gain structure within a digital mixer must be such that there is an appropriate dynamic window for the signals at each point in the chain, also allowing for operations such as equalisation which change the signal level.

2.8.4 Digital filters and equalisation

Digital filtering is something of a 'catch-all' term, and is often used to describe DSP operations which do not at first sight appear to be filtering. A digital filter is essentially a process which involves the time delay, multiplication and recombination of audio samples in all sorts of configurations, from the simplest to the most complex. Using digital filters one can create low- and

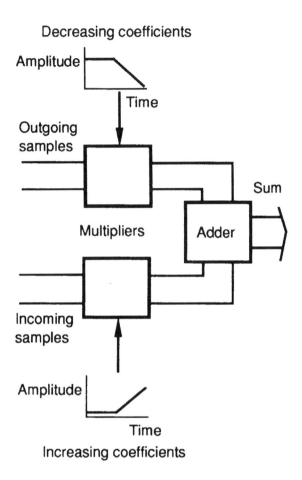

Figure 2.33 Conceptual block diagram of the crossfading process, showing two audio signals multiplied by changing coefficients, after which they are added together

high-pass filters, peaking and shelving filters, echo and reverberation effects, and even adaptive filters which adjust their characteristics to affect different parts of the signal.

To understand the basic principle of digital filters it helps to think about how one might emulate a certain analog filtering process digitally. Filter responses can be modelled in two main ways – one by looking at their frequency domain response and the other by looking at their time domain response. (There is another approach involving the so-called z-plane transform, but this is not covered here.) The frequency domain response shows how the amplitude of the filter's output varies with frequency, whereas the time domain response is usually represented in terms of an impulse response (see Figure 2.34). An impulse response shows how the filter's output responds to stimulation at the input by a single short impulse. Every frequency response

Figure 2.34 Examples of (a) the frequency response of a simple filter, and (b) the equivalent time domain impulse response

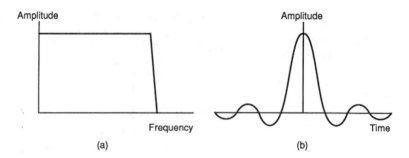

has a corresponding impulse response because the two are directly related. If you change the way a filter responds in time you also change the way it responds in frequency. A mathematical process known as the Fourier transform is often used as a means of transforming a time domain response into its equivalent frequency domain response. They are simply two ways of looking at the same thing.

Digital audio is time discrete because it is sampled. Each sample represents the amplitude of the sound wave at a certain point in time. It is therefore normal to create certain filtering characteristics digitally by operating on the audio samples in the time domain. In fact if it was desired to emulate a certain analog filter characteristic digitally one would need only to measure its impulse response and model this in the digital domain. The digital version would then have the same frequency response as the analog version, and one can even envisage the possibility of favourite analog filters being recreated for the digital workstation. The question, though, is how to create a particular impulse response characteristic digitally, and how to combine this with the audio data.

As mentioned earlier, all digital filters involve delay, multiplication and recombination of audio samples, and it is the arrangement of these elements that gives a filter its impulse response. A simple filter model is the finite impulse response (FIR) filter, or transversal filter, shown in Figure 2.35. As can be seen, this filter consists of a tapped delay line with each tap being multiplied by a certain coefficient before being summed with the outputs of the other taps. Each delay stage is normally a one sample period delay. An impulse arriving at the input would result in a number of separate versions of the impulse being summed at the output, each with a different amplitude. It is called a finite impulse response filter because a single impulse at the input results in a finite output sequence determined by the number of taps. The more taps there are the more intricate

Figure 2.35 A simple FIR filter (transversal filter). N = multiplication coefficient for each tap. Response shown below indicates successive output samples multiplied by decreasing coefficients

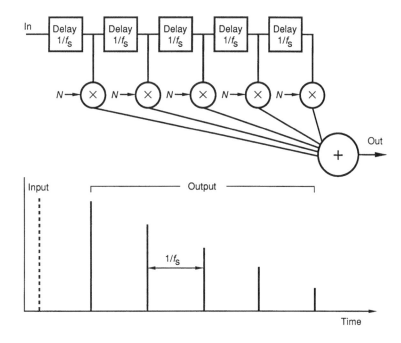

Figure 2.36 A simple IIR filter (recursive filter). The output impulses continue indefinitely but become very small. N in this case is about 0.8. A similar response to the previous FIR filter is achieved but with fewer stages

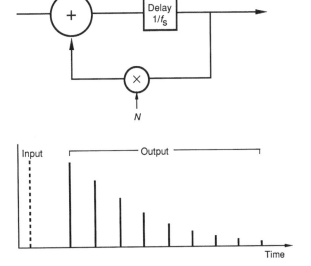

the filter's response can be made, although a simple low-pass filter only requires a few taps.

The other main type is the infinite impulse response (IIR) filter which is also known as a recursive filter, because there is a degree of feedback between the output and the input (see Figure 2.36). The response of such a filter to a single impulse is an infinite output sequence, because of the feedback. IIR filters are often used in audio equipment because they involve fewer elements for most variable equalisers than equivalent FIR filters, and they are useful in effects devices. They are unfortunately not phase linear, though, whereas FIR filters can be made phase linear.

2.8.5 Digital reverberation and other effects

It can probably be seen that the IIR filter described in the previous section forms the basis for certain digital effects, such as reverberation. The impulse response of a typical room looks something like Figure 2.37, that is an initial direct arrival of sound from the source, followed by a series of early reflections, followed by a diffuse 'tail' of densely packed reflections decaying gradually to almost nothing. Using a number of IIR filters, perhaps together with a few FIR filters, one could create a suitable pattern of delayed and attenuated versions of the original impulse to simulate the decay pattern of a room. By modifying the delays and amplitudes of the early reflections and the nature of the diffuse tail one could simulate different rooms.

The design of convincing reverberation algorithms is a skilled task, and the difference between crude approaches and good ones is very noticeable. Some audio workstations offer limited reverberation effects built into the basic software package, but these often sound rather poor because of the limited DSP power available and the crude algorithms involved. More convincing reverberation processors are available which exist either as stand-alone devices or as optional plug-in cards to the workstation, having more DSP capacity and tailor-made software.

Figure 2.37 The impulse response of a typical reflective room

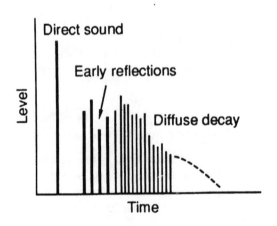

Other simple effects can be introduced without much DSP capacity, such as double-tracking and phasing/flanging effects. These often only involve very simple delaying and recombination processes. Pitch shifting can also be implemented digitally, and this involves processes similar to sample rate conversion, as described below. High quality pitch shifting requires quite considerable horsepower because of the number of calculations required.

2.8.6 Dynamics processing

Digital dynamics processing involves gain control which depends on the instantaneous level of the audio signal. A simple block diagram of such a device is shown in Figure 2.38. A side chain produces coefficients corresponding to the instantaneous gain change required, which are then used to multiply the delayed audio samples. First the r.m.s. level of the signal must be determined, after which it needs to be converted to a logarithmic value in order to determine the level change in decibels. Only samples above a certain threshold level will be affected, so a constant factor must be added to the values obtained, after which they are multiplied by a factor to represent the compression slope. The coefficient values are then antilogged to produce linear coefficients by which the audio samples can be multiplied.

Figure 2.38 A simple digital dynamics processing operation

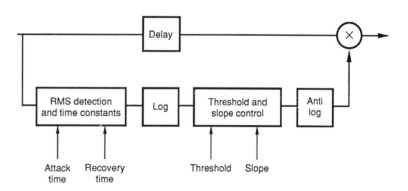

2.8.7 Sample rate conversion

Sample rate conversion is necessary whenever audio is to be transferred between systems operating at different rates. The aim is to convert the audio to the new rate without any change in pitch or addition of distortion or noise. These days sample rate conversion can be a very high quality process, although it is never an entirely transparent process because it involves modifying the sample values and timings. As with requantising algorithms, it is fairly common to encounter poorly implemented sample rate conversion on low cost digital audio workstations, often depending very much on the specific software application rather than the hardware involved.

The easiest way to convert from one rate to another is by passing through the analog domain and resampling at the new rate, but this may introduce a small amount of extra noise. The most basic form of digital rate conversion involves the translation of samples at one fixed rate to a new fixed rate, related by a simple fractional ratio. This was the reason for the suggestion that 50.4 kHz should be the new professional sampling rate some years ago, since it was related to 44.1 in the ratio 8/7. Latterly, 48 kHz was chosen as the professional recording rate for broadcast use because it has a simple fractional relationship to 32 kHz.

Fractional-ratio conversion involves the mathematical calculation of samples at the new rate based on the values of samples at the old rate. Digital filtering is used to calculate the amplitudes of the new samples such that they are correct based on the impulse response of original samples, after low-pass filtering with an upper limit of the Nyquist frequency of the original sampling rate. A clock rate common to both sample rates is used to control the interpolation process. Using this method, some output samples will coincide with input samples, but only a limited number of possibilities exist for the interval between input and output samples.

If the input and output sampling rates have a variable or non-simple relationship the above does not hold true, since output samples may be required at any interval in between input samples. This requires an interpolator with many more clock phases than for fractional-ratio conversion, the intention being to pick a clock phase which most closely corresponds to the desired output sample instant at which to calculate the necessary coefficient. There will clearly be an error, which may be made smaller by increasing the number of possible interpolator phases. The audible result of the timing error is equivalent to the effects of jitter on an audio signal (see above), and should be minimised in design so that the effects of sample rate conversion are below the noise floor of the signal resolution in hand. If the input sampling rate is continuously varied (as it might be in variable-speed searching or cueing), the position of interpolated samples with relation to original samples must vary also, and this requires real-time calculation of filter phase.

Many workstations now include sample rate conversion as either a standard or optional feature, so that audio material recorded and edited at one rate can be output at another (e.g. 48 kHz DAT recordings, edited and transferred to a CD master at 44.1 kHz, or 44.1 kHz material converted to 22 kHz for multimedia applications). It is important to ensure that the quality of the sample rate conversion is high enough not to affect the sound quality of your recordings, and it should only be used if it cannot be avoided. Poorly implemented applications sometimes omit to use correct low-pass filtering to avoid aliasing, or incorporate very basic digital filters, resulting in poor sound quality after rate conversion.

Sample rate conversion is also useful as a means of synchronising an external digital source to a standard sampling frequency reference, when it is outside the range receivable by a workstation, as introduced in Chapter 6.

References

1   Shannon, C. E. (1948) 'A mathematical theory of communication'. *Bell System Technical Journal*, **27**, p. 379.
2   Vanderkooy, J. and Lipshitz, S. (1989) 'Digital dither'. In *Audio in Digital Times*, Audio Engineering Society, New York.
3   Cohen, E. and Fielder, L. (1992) 'Determining noise criteria for recording environments'. *Journal of the Audio Engineering Society*, **40**, 5, p. 384.
4   Fielder, L. (1995) 'Dynamic range issues in the modern digital audio environment'. *Journal of the Audio Engineering Society*, **43**, 5, pp. 322–339.
5   Ohashi, T. *et al* .(1991) 'High frequency sound above the audible range affects brain electrical activity and sound perception'. Presented at 91st AES Convention, New York, preprint 3207.
6   Dunn, N. J. (1992) 'Jitter: specification and assessment in digital audio equipment'. Presented at the 93rd AES Convention, San Francisco, 1–4 October, preprint no.3361 (C-2).
7   Dunn, N. J. (1991) 'Considerations for interfacing digital audio equipment to the standards AES3, AES5 and AES11'. In Proceedings of the AES 10th International Conference, 7–9 September, pp. 115–126.
8   Story, M. (1989) 'The AES interface and analog–digital conversion'. In Proceedings of the AES/EBU Interface Conference, 12–13 September, London, pp. 16–21, Audio Engineering Society British Section.
9   Dunn, C. and Hawksford, M. O. (1992) 'Is the AES/EBU/SPDIF digital audio interface flawed?' Presented at the 93rd AES Convention, San Francisco, 1–4 October, preprint no. 3360 (C-1).

10   Stuart, J. R. and Wilson, R. J. (1994) 'Dynamic range enhancement using noise shaped dither applied to signals with and without preemphasis'. Presented at the 94th AES Convention, Amsterdam, 26 February – 1 March, preprint no. 3871 (P7.1).

Recommended further reading

Pohlmann, K. (1995) *Principles of Digital Audio*. McGraw Hill.
Watkinson, J. (1994) *An Introduction to Digital Audio*. Focal Press.

3 Audio data reduction systems

It is common now to encounter data reduction in digital audio. This is the coding of audio at lower bit rates than those required by linear PCM, using processing in the digital domain which removes irrelevant or redundant information from the bit stream (see Figure 3.1). The aim is to preserve as much as possible of the sound quality whilst reducing the bit rate. Using such techniques it is possible to reduce the bit rate quite significantly so that longer storage times and more simultaneous channels of replay may be obtained from disks and other media. In many cases, though, data reduction involves a trade-off between bit rate savings and sound quality which must be carefully judged. It is likely that the process will be used quite widely in multimedia applications, especially in the distribution of the end product, but it is less likely that it will be used for high quality audio origination and post production work.

Figure 3.1 Audio data reduction and subsequent reconstruction

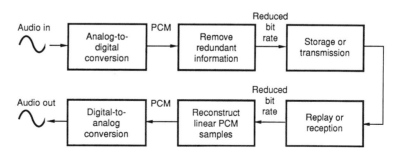

This chapter will explain how audio data reduction (sometimes called 'compression') works, and will describe the most common standards in the field, including the ISO/MPEG standard. It will also highlight the sound quality issues at stake, and provide practical advice on the correct use of data reduction.

3.1 Why reduce the data rate?

Nothing is inherently wrong with linear PCM from a sound quality point-of-view, indeed it is the best thing to use. The problem is simply that the data rate is too high for a number of applications. Two channels of linear PCM require a rate of around 1.4 Mbit/s, whereas applications such as Digital Audio Broadcasting and ISDN communications need it to be more like 128 kbit/s – in other words more than ten times less data per second. We know that with linear PCM the sampling rate dictates audio bandwidth, and that the number of bits per sample dictates signal-to-noise ratio. These are also the factors that dictate the resulting data rate, and neither parameter can usefully be reduced much for a broadband digital signal without a noticeable loss of sound quality.

High data rates equate with high cost – either of storage, transfer or transmission – and thus a considerable amount of work has been done over the past few years to develop systems capable of reducing data rate without reducing sound quality. Data reduction is not a new phenomenon, since systems based on what now seem to be primitive methods have been in operation in broadcasting and telecommunications for years, but the current interest centres on what are commonly called 'perceptual coders' which rely on the use of a detailed model of the human hearing mechanism to decide what is perceptually 'redundant' in the audio signal. These systems offer previously impossible degrees of reduction in data rate, and operate in real time with a short coding delay. They are possible because the speed and cost of digital signal processing have now reached a point at which the required operations can be performed cost effectively.

As explained in Chapters 4 and 6, the efficiency of mass storage media and data networks is related to their data transfer rates. The more data can be moved per second, the more audio channels may be handled simultaneously, the faster a disk can be copied, the faster a sound file can be transmitted across the world. In reducing the data rate that each audio channel demands, one also reduces the requirement for such high performance from storage media and networks, or alternatively one can obtain greater functionality from the same performance as before. A disk drive capable of replaying eight channels of linear PCM simultaneously could be made to replay, say, 48 channels of data-reduced audio, without unduly affecting sound quality.

Although this sounds like magic, and makes it seem as if there is no point in continuing to use linear PCM, it must be appreciated that the data reduction is achieved by throwing away data from the original audio signal. The more data is thrown away the more likely it is that unwanted audible effects will be noticed. The design aim of most of these systems is to try to retain as much as possible of the sound quality whilst throwing away as much data as possible, so it follows that one should always use the least data reduction necessary, where there is a choice.

3.2 Lossless and lossy coding

There is an important distinction to be made between the type of data reduction used in some computer applications and the approach used in many audio coders. The distinction is really between 'lossless' coding and coding which involves some loss of information (see Figure 3.2). It is quite common to use data compression on computer files in order to fit more information onto a given disk or tape, but such compression is usually lossless in that the original data is reconstructed bit for bit when the file is decompressed. A number of tape backup devices for computers have a compression facility for increasing the apparent capacity of the medium, for example. Methods are used which exploit redundancy in the information, such as coding a string of eighty zeros by replacing them with a short message stating the value of the following data and the number of bytes

Figure 3.2 (a) In lossless coding the original data is reconstructed perfectly upon decoding, resulting in no loss of information. (b) In lossy coding the decoded information is not the same as that originally coded, but the coder is designed so that the effects of the process are minimal

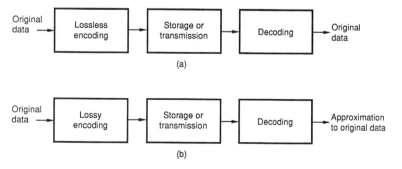

75

involved. This is particularly relevant in single-frame bit-mapped picture files where there may be considerable runs of black or white in each line of a scan, where nothing in the image is changing. One may expect files compressed using off-the-shelf PC data compression applications to be reduced to perhaps 25–50 per cent of their original size, but it must be remembered that they are dealing with static data, and do not have to work in real time. Also, it is not normally acceptable for decompressed computer data to be anything *but* the original data.

It is possible to use lossless coding on audio signals, and an example is described by Cellier.[1] Lossless coding allows the original PCM data to be reconstructed perfectly by the decoder, and is therefore 'noiseless' since there is no effect on audio quality. The data reduction obtained using these methods ranges from nothing to about 2.5:1, and is variable depending on the programme material. This is because audio signals have an unpredictable content, do not make use of a standard character set, and do not spend long periods of time in one binary state or the other. Although it is possible to perform this reduction in real time, the coding gains are not sufficient for many applications. None the less, a halving in the average audio data rate is certainly a useful saving, and there are a growing number of professional disk-based audio systems which make use of it.

'Noisy' or lossy coding methods make possible a far greater degree of data reduction, but require the designer and user to arrive at a compromise between the degree of data reduction and potential effects on sound quality. Here data reduction is achieved by coding the signal less accurately than in the original PCM format (using fewer bits per sample), thereby increasing quantising noise, but with the intention that increases in noise will be 'masked' by the signal. The original data is not reconstructed perfectly on decoding. The success of such techniques therefore relies on being able to model the characteristics of the human hearing process in order to predict the masking effect of the signal at any point in time – hence the common term 'perceptual coding' for this approach. Using detailed psychoacoustic models it is possible to code high quality audio at rates under 100 kbit/s per channel with minimal effects on audio quality. Higher data rates, such as 192 kbit/s, can be used to obtain an audio quality which is demonstrably indistinguishable from the original PCM.

The majority of the following chapter deals with perceptual coding schemes, since these form the basis for MPEG-Audio, which is an internationally standardised scheme.

3.3 Psychoacoustic principles of lossy audio data reduction

In order to understand perceptual audio coding it is necessary to introduce a small amount of psychoacoustics – the study of auditory perception. Masking occurs in auditory perception when one sound causes another to be either less audible or inaudible. Also one component of a complex sound can mask other less prominent components. The principles of masking are widely documented in the literature, and it is only proposed to give a summary here. A useful discussion may be found in Moore.[2]

Masking occurs in both the time and frequency domains. Simultaneous masking occurs when one sound is masked by another which is sounded at the same time. The masking effect of a pure tone extends slightly below and some way above its own frequency, depending on its level, as shown in the example of Figure 3.3 which plots hearing threshold against sound pressure level. Sounds whose level is below the masking curve will not normally be perceived, although there are instances in which such sounds may be 'unmasked' by others coming from different directions, as described by Gerzon.[3] The greatest masking effect occurs when two signals occupy the same critical band within the auditory system. Critical bands were first proposed by Fletcher [4] in the 1940s, and the concept was refined, particularly by Zwicker,[5] in the 1960s. They have been determined by Zwicker as a number of separate frequency bands (24 between 20 Hz and 15 kHz) within which the masking effect is constant and independent of the separation in frequency between the 'masking' and the 'masked' sound. Critical bands represent the initial 'spectrum analysis' performed by the

Figure 3.3 Example of the masking effect of a pure tone at 400 Hz and 80 dB SPL. The shape and extent of the masking effect depends on the frequency of the tone and its level. The masking effect of tones is different from that of noise

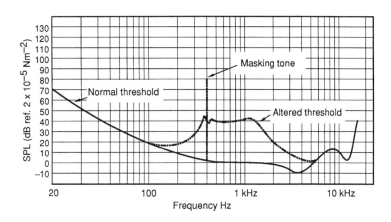

Figure 3.4 Simultaneous masking occurs whilst the masker is sounding. The backward masking effect is much less extended in time than the forward masking effect

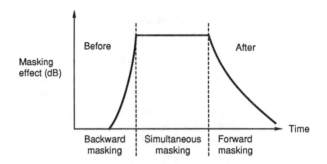

hearing mechanism. Once two sounds become separated in frequency by more than a critical band, the masking effect of one on the other is reduced, (although it may still be present, especially above the frequency of the louder signal). Sources differ as to the precise width of the critical band, but it is considered to be between one-third and one-fifth of an octave at middle to high frequencies, and considerably wider at low frequencies.

Non-simultaneous masking is divided into backward and forward masking, as illustrated in Figure 3.4. In forward masking a sound may mask another which occurs a short time after it, and in backward masking the effect *precedes* the sound. The extent of forward masking depends on the level and nature of the signal, but may extend up to 100 ms or so, whilst backward masking is less easy to demonstrate and only extends up to roughly 5 ms before the sound. Some trained listeners do not notice the backward masking effect at all.

3.4 General principles of a perceptual coder

With ordinary PCM there are only two ways to reduce the data rate: either reduce the audio bandwidth (use a lower sampling rate) or allow an increase in the noise level (use fewer bits per sample). Both of these normally make the sound quality poorer. Perceptual coding exploits the masking phenomenon, allowing quantising noise to be increased considerably without affecting sound quality. In fact quantising noise can be allowed to rise by amounts previously considered totally unreasonable, provided that it remains below the masking threshold of the audio signal concerned.

Perceptual coders work by splitting the audio signal into a number of narrow bands and then requantising each band using fewer bits, under the control of a psychoacoustic model, as shown in Figure 3.5. The band-splitting function is achieved in one of two ways: either using a digital filterbank to create a

Figure 3.5 Generalised block diagram of a psychoacoustic low bit rate coder

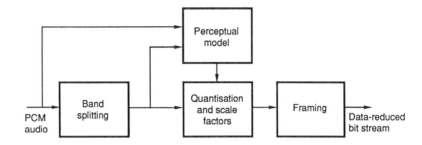

number of sub-bands, or using a mathematical transform such as the modified discrete cosine transform (MDCT) to create a number of frequency-domain coefficients from a block of audio samples. The psychoacoustic model analyses short blocks of audio samples (usually a few milliseconds at a time) to determine their frequency spectrum (the level of audio at each frequency), and calculates the overall masking threshold which applies to that block. This is then used to control the requantisation of the audio samples in each frequency band, according to the degree of masking which affects each band.

As explained in Chapter 2, coarser quantising results in increased quantising noise and thus would normally be unacceptable, but here requantisation effects are constrained within narrow bands and thus are more effectively masked than if the noise resulting from requantisation were to be spread across the whole frequency range. As shown in Figure 3.6, noise in the same band as a signal component can be at quite a high level and still lie under the masking threshold. The coder allocates bits to each filtered band, giving more bits to those bands where not much masking takes place, and fewer bits to those which experience a lot of masking. It is even possible not to code some bands at all if they are totally masked.

Figure 3.6 Quantising noise lying under the masking threshold will normally be inaudible

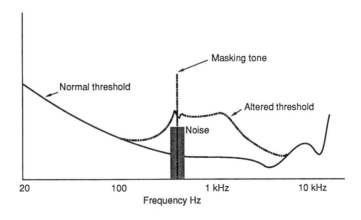

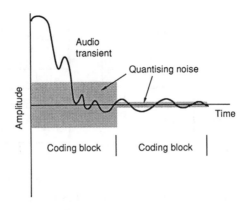

Figure 3.7 At sharp audio transients noise may be audible during the quiet period either before or after the transient. This is because the masking effect is calculated over a block of samples lasting a number of milliseconds. Here it can be seen that the constant noise level during the first block becomes unmasked in the latter half of the block, as the signal level falls. In the next block the noise level is much lower

Because the audio is analysed in blocks, the masking threshold calculated by the model is basically a form of average threshold for the block. If the block contains some very loud audio followed by some very quiet audio (such as at a transient) then it is possible that noise which is masked in the loud part will be above the threshold in the quiet part (see Figure 3.7). Backward and forward masking effects come into play here, because they may help to hide the short parts of a block when noise would otherwise be audible. Some systems employ dynamic window switching techniques to vary the length of the coding block depending on the transient content of the signal, in an attempt to minimise this effect.

The only mainstream lossy audio data reduction system that does not use perceptual coding at the moment is apt-X 100,[6] and this relies principally on techniques developed from speech coding such as adaptive differential pulse code modulation (ADPCM) and linear predictive coding (LPC) to achieve a data reduction ratio of 4:1.

3.5 ISO-MPEG audio data reduction

3.5.1 Overview

As a means of exploring the principles involved in a typical low bit rate audio system, the MPEG-Audio standard will be described in detail. Features of other systems will only be summarised.

The ISO standard 11172[7] describes coding of moving pictures and associated audio for digital storage media at up to about 1.5 Mbit/s. This is the so-called MPEG-1 standard. Four different audio channel modes are possible: single channel, dual channel (two independent channels in one bitstream), stereo and

Table 3.1 MPEG-Audio layers

Layer	Complexity	Min. delay	Bit rate range	Target
1	Low	19 ms	32–448 kbit/s	192 kbit/s
2	Moderate	35 ms	32–384 kbit/s*	128 kbit/s
3	High	59 ms	32–320 kbit/s	64 kbit/s

* In Layer 2, bit rates of 224 kbit/s and above are for stereo modes only

joint stereo (stereo coding which takes advantage of left/right channel redundancy).

Three 'layers' of complexity have been specified for audio data reduction, ranging from Layer 1 (the least complex) to Layer 3. Total data rates of between 32 and 448 kbit/s are accommodated, depending on the Layer and mode (see Table 3.1). The higher-numbered layers, being more complex, obtain better audio quality at lower data rates, but with a longer coding delay. Theoretical minima for encoding/decoding delay are roughly Layer 1: 19 ms; Layer 2: 35 ms; Layer 3: 59 ms. Consequently, Layer 1 is most appropriate for applications where high audio quality is required with minimal delay, and where higher data rates such as 192 kbit/s are available. Layers 2 and 3 are more appropriate at rates of 128 kbit/s and below. The target bit rates do not restrict a scheme to operating only at one bit rate. For example, recent developments in Layer 2 coding have shown that it can operate at lower data rates with good results.

Some confusion has arisen over the names of audio data reduction systems which have formed parts of the ISO standard. People may still refer to the systems by name rather than ISO layer. MUSICAM is the system upon which Layers 1 and 2 was based, and when people refer to a MUSICAM encoder then they almost certainly mean an ISO-MPEG Layer 2 encoder. There may be older MUSICAM coders in existence which do not conform to the letter of the ISO standard. ASPEC is a system which forms part of the Layer 3 specification.

The ISO standard defines only the format of the encoded data stream and the decoding process, leaving the way clear for improvements to be made to the psychoacoustic model used in the encoder, if and when they become possible. None the less, quite detailed guidance is given in the standard concerning encoding and psychoacoustic models. No psychoacoustic model is required in the decoder, since the decoder simply acts to 'unpack' the audio data and scalefactors contained in the encoded bit stream in order to reconstruct PCM audio samples

Figure 3.8 Block diagrams of MPEG-Audio encoders, after Brandenburg and Stoll. (a) Layer 1; (b) Layer 2; (c) Layer 3

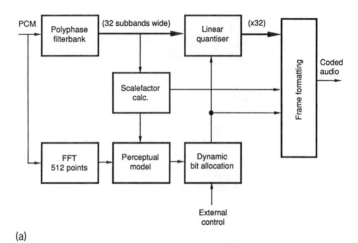

(a)

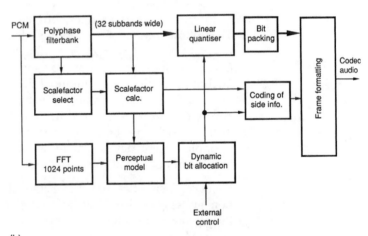

(b)

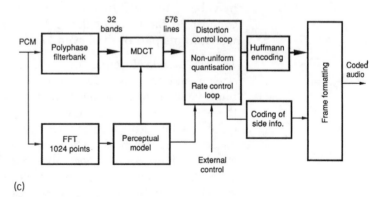

(c)

at the original rate. A decoder can therefore be made more cheaply than an encoder.

3.5.2 Encoding system

In all three layers of the ISO scheme the audio data is initially split into 32 subbands using a digital filter. An additional filter bank is used in Layer 3 to increase the frequency resolution up to 576 independent bands, and a different psychoacoustic analyser is used to improve the modelling of the hearing process. Block diagrams of the encoding processes are shown in Figure 3.8.

In Layer 1, blocks of 384 PCM samples are divided into 32 subbands, resulting in 12 subband samples per subband. The samples in the block are then scaled based on the highest level sample in each group of 12 subband samples. This has the effect of removing leading zeros (or ones for negative values) from low level samples, shifting the active bits into the MSB positions. The scaled samples are then linearly requantised according to the number of bits available by removing LSBs and rounding appropriately. Bits are allocated to each subband according to their need (from a sound quality point of view, based on the masking model's calculation of the masking level in each subband) and the required output bit rate. Between 0 and 15 bits per subband sample are possible.

The audio data part of each frame consists of bit allocation information, scalefactor information and subband sample data, as shown in Figure 3.9(a) and (b). The bit allocation data indicates, for each subband, how many bits per sample have been allocated. The 6 bit scale factor indicates, for each subband, the factor by which the requantised samples must be multiplied at the decoder in order to restore them to their correct level.

Figure 3.9 MPEG-Audio frame format

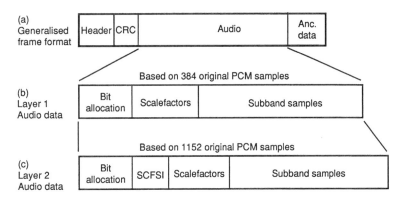

In Layer 2, a certain improvement in economy is achieved by grouping sample data, and also by grouping scale factors and bit allocation information so that they apply to more than one group of 12 subband samples. The Layer 2 frame (see Figure 3.9(c)) represents three times the number of PCM samples contained in a Layer 1 frame, but not necessarily three times the number of scale factors or audio sample values. Consequently the Layer 2 frame also contains 'scale factor select information' (SCFSI) which indicates how many scale factors have been included for each subband. To take an example, it might be possible for only one scale factor to be transmitted, applying to all three groups of samples in a particular subband. This could happen if the maximum signal level in three successive groups of samples was sufficiently similar. Subband samples themselves may also be grouped into single codewords representing three consecutive samples, and this is indicated by the bit allocation information, related to a table contained in the decoder.

The additional complexity of Layer 3 is quite considerable, and it is not proposed to describe in detail the various methods by which Layer 3 indicates bit allocation in the frame format. Suffice it to say that in order to achieve good sound quality at low bit rates Layer 3 additionally uses a noise-shaping quantiser, dynamic window switching (which optimises the length of blocks depending on the transient content of the audio), high-efficiency Huffmann encoding (a lossless 'packing' technique) and a 'bit reservoir' technique which improves the coding of critical blocks.

3.5.3 Frame format

The frame format for MPEG data-reduced audio was shown in Figure 3.9. Layer 1 frames correspond to 384 original PCM samples (8 ms at a sampling rate of 48 kHz), and Layer 2 and 3 frames correspond to 1152 PCM samples (24 ms at 48 kHz). The frame consists of a 32 bit header, a 16 bit CRC (cyclic redundancy check) check word, the audio data (consisting of subband samples, appropriate scale factors and information concerning the bit allocation to different parts of the spectrum) and an ancillary data field whose length is currently unspecified.

The 32 bit header of each frame consists of the following information:

Sync word (12 bits) — All set to binary '1' to act as a synchronisation pattern

ID bit (1 bit) — Indicates the ID of the algorithm in use (set to '1' for ISO 11172-3)

Layer (2 bits)	– Indicates the layer in use: '11' = Layer 1; '10' = Layer 2; '01' = Layer 3
Protection bit (1 bit)	– Indicates whether error correction data has been added to the audio bitstream ('0' if yes)
Bitrate index (4 bits)	– Indicates the total bit rate of the channel according to a table which relates the state of these 4 bits to rates in each layer
Sampling frequency (2 bits)	– Indicates the original PCM sampling frequency: '00' = 44.1 kHz; '01' = 48 kHz; '10' = 32 kHz
Padding bit (1 bit)	– Indicates in state '1' that a slot has been added to the frame to make the average bit rate of the data reduced channel relate exactly to the original sampling rate
Private bit (1 bit)	– Available for private use
Mode (2 bits)	– '00' = stereo; '01' = joint stereo; '10' = dual channel; '11' = single channel
Mode extension (2 bits)	– used for further definition of joint stereo coding mode to indicate either which bands are coded in joint stereo, or which type of joint coding is to be used
Copyright (1 bit)	– '1' = copyright protected
Original/copy (1 bit)	– '1' = original; '0' = copy
Emphasis (2 bits)	– Indicates audio pre-emphasis type: '00' = none; '01' = 50/15µs; '11' = CCITT J17.

3.5.4 Decoding

The decoding of the data-reduced audio information is basically the reverse of the encoding process, without the psychoacoustic model, as shown in Figure 3.10. PCM samples are reconstructed from the subband samples using the scale factors and bit allocation information, followed by resynthesis of the original audio band from the subbands using a suitable synthesis filter.

Figure 3.10 Generalised block diagram of an MPEG-Audio decoder

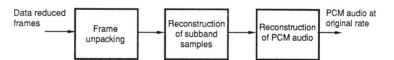

85

3.5.5 *Joint stereo coding*

One means of improving the sound quality at low data rates is to use so-called 'joint stereo coding' techniques. As the name implies, these are only effective when used with stereo pair signals, where it may be possible to obtain some coding gain by exploiting commonality between the channels. Two main methods are employed: one using sum and difference or MS (main–side) stereo, and the other using intensity stereo. The MS method is only available for use with Layer 3, whilst the intensity method may be used in any layer.

MS processing involves the conversion of the stereo signal from left–right format to sum and difference format. M is the sum of left and right channels (L+R), whilst S is the difference between them (L–R). In the case of most real stereo signals the M signal is considerably higher in level than the S signal, and the closer a signal is to being monophonic the greater the difference between M and S. If a near-monophonic signal with stereo reverberation is data-reduced in the LR format, the quantising noise will be different in the two channels. On replay, the mono signal will appear to be coming from the centre, whereas the noise will be spread across the stereo sound stage, thus making it easier to perceive. If the same signal is coded in the MS format, then most of the coding noise will lie in the M channel, and will be split equally and coherently between left and right on decoding, thus placing the noise in the same stereo location as the signal. It is more likely to be masked by the signal in such a situation.

The decision over whether to employ MS coding as opposed to LR coding is based on a comparison in the encoder of the likely number of bits required for the two schemes. Savings in bit rate may be obtained by setting the S component to zero for certain subbands in which its contribution to the stereo signal is judged to be irrelevant. The threshold of this judgement is set so as not to affect the stereo impression on decoding.

Intensity stereo is based on the principle that the hearing process becomes gradually poorer at detecting source position at high frequencies (above about 2 kHz), and that detection of position at HF is based almost entirely on intensity difference between the channels. Consequently, above a certain frequency, selected according to best judgement, the intensity stereo method may be used to transmit only one channel of subband sample information (being the sum of left and right samples) and one set of bit allocation data, whilst still sending the scale factor information for both channels. This approach is used in Layers 1 and 2. In Layer 3, it is possible to send a single scale factor with the monophonic subband sample data, along with a second value indicating the stereo image position for each band. This position value alters the ratio between left and right levels for the subband on decoding.

3.5.6 Data rate versus sound quality

It is hard to make absolute statements concerning the relationship between data rate and sound quality in MPEG systems. Since improvements in the psychoacoustic model used in the coder can be made, and indeed are being made all the time, the situation changes as time passes. In general, though, for sampling rates of 48 kHz it would be fair to say that very high audio quality is achievable from Layer 1 and 2 coders operating at data rates higher than 128 kbit/s per channel. Below 100 kbit/s one increasingly notices artefacts, but they are highly dependent on the type of programme material. At 64 kbit/s per channel the quality ranges from acceptable to quite poor. Appropriate use of joint stereo coding improves the situation for stereo signals by an important degree, such that at 2x64 kbit/s a stereo signal sounds better than a mono signal at 64 kbit/s. Although the target data rate for Layer 2 was supposed to have been 128 kbit/s per channel, recent development work has brought its performance close to that of Layer 3 when using joint stereo coding at an overall bit rate of around 128 kbit/s.[8]

3.6 MPEG 2 – second phase enhancements and additions

A second phase of work, informally known as MPEG 2 and described in ISO 13818,[9] has been directed at introducing compatible enhancements to the original ISO standard. These include dynamic range control, multichannel sound and the use of lower sampling rates.

3.6.1 MPEG-2 dynamic range control

It is very easy to implement compression of the dynamic range of audio programme material in an MPEG decoder. Compression may be used to match the dynamic range of the programme to the listening environment, in order that digital radio programmes might be listened to comfortably in the car, for example. It is proposed that Layer 1 and 2 MPEG-2 decoders should allow for dynamic range control of this type by making it possible for the scale factors to be multiplied by an appropriate gain change factor, coupled with suitable time constants to avoid the typical overshoot and modulation effects of all forms of compression. Experiments have shown that high quality dynamic range control may be implemented with little additional decoder complexity.

3.6.2 Multichannel sound

The extension of the ISO-MPEG coding scheme to multichannel sound, in a fashion compatible with MPEG-1 decoders, is the main item of importance in MPEG-2. Multichannel sound is gaining increased importance, especially with relation to future standards for HDTV (high definition television), and it is likely that broadcasts of HDTV programmes in the future will be accompanied by surround sound. There is also potential for audio-only uses of surround sound. The SMPTE, CCIR and EBU have all published draft standards for multichannel sound formats, with or without pictures, involving 5 channels, with an optional subwoofer channel. The 5 channels break down into three front channels and two surround channels, making it a so-called 3/2 surround format.

The surround option within MPEG-2 is designed to transmit the five channels, data reduced, in a discrete fashion, using a compatibility matrix to allow the basic stereo left and right channels to be transmitted in the part of the frame used by MPEG-1 decoders.[10,11] Figure 3.11 shows the frame format devised for surround information. The multichannel extension data is contained within that part of the frame set aside for ancillary data in the ISO 11172-3 standard (MPEG-1). Using the equivalent of joint stereo coding for multichannel information, the combined bit rate for the 5 channel surround data will be between 256 and 384 kbit/s, when using Layer 2.

Figure 3.11 Frame format for MPEG 2 multichannel extension

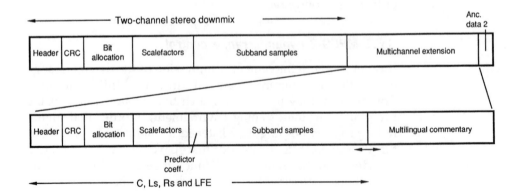

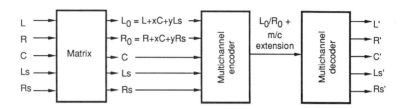

Figure 3.12 Compatibility matrixing for MPEG 2 multichannel extension

Figure 3.12 shows how the 5 channels of information are derived in a compatible manner. Compatible two-channel stereo signals are obtained from a downmix of the surround information, forming L_0 and R_0 signals, to be transmitted in the main part of the frame. The centre and stereo surround signals are transmitted separately in the multichannel extension part of the frame. On reception by a multichannel decoder, the original surround channels are reconstituted by combining the signals in an appropriate matrix. Since the system is digital, the matrixing process will not suffer from the phase and amplitude errors which can distort equivalent analog processes.

It has also been proposed that the multichannel extension bitstream will carry up to 7 multilingual channels, and a low frequency effects (subwoofer) channel.

3.6.3 Lower sampling frequencies

Another of the key aims of MPEG-2 has been to extend the ISO standard to lower data rates and sampling frequencies. This would allow the use of reasonably high quality audio in such applications as those involving multimedia computers, ISDN, commentary and talkback, whilst allowing the use of data rates between 24 and 64 kbit/s per channel. This would offer an attractive alternative to older coding methods such as G722.

MPEG-2 contains three new audio sampling frequencies of exactly half the primary rates, i.e.: 24, 22.05 and 16 kHz. The use of these lower sampling frequencies results in narrower subbands, and thus a further constraining of quantising noise. Demonstrations of MPEG-2, Layer 2 at 48 kbit/s with speech show that speech sampled at 24 kHz sounds quite acceptable for commentary purposes, and suffers little from the loss of frequencies above 12 kHz.

3.7 Non-MPEG audio data reduction systems

The following is a resumé of audio data reduction systems that are not part of the MPEG standard. They are used in a number of commercial systems worldwide, and may be the subject of licence agreements.

3.7.1 PASC

PASC (Precision Adaptive Sub-band Coding) is the system used in Philips' DCC (the recently developed consumer cassette system). It is almost identical to MUSICAM, the system on which Layers 1 and 2 of the ISO/MPEG standard are based.

3.7.2 ATRAC

ATRAC (Adaptive Transform Acoustic Coding) is the system used for Sony's MiniDisc and in the Sony Dynamic Digital Sound (SDDS) multichannel cinema system. ATRAC is a transform coder offering data reduction to around one fifth of the original rate.

3.7.3 Dolby AC-1, AC-2 and AC-3

AC-1 is Dolby's original adaptive delta modulation process, stemming from developments in the early eighties, operating at a data rate of just over 200 kbit/s per channel. AC-2 is Dolby's second generation family of low bit rate transform coding algorithms. The company manufactures both a 'moderate delay' coder (55 ms) and a 'low delay' coder in the AC-2 family, with an output data rate of 128 kbit/s per channel at a 48 kHz audio sampling rate. The system is also capable of being extended to lower data rates.

AC-3 is the name given to the data reduction process used, amongst other things, for multichannel sound in the Dolby SR-D system for digital sound tracks on cinema film, and for multichannel sound on some video Laser Discs. It uses a data rate for 5.1 channels of audio not substantially greater than that used for stereo in AC-2, by exploiting similarities between the channels and using additional techniques to reduce the data rate of the scaling factors. It is gaining considerable favour for broadcasting and other applications which may involve multichannel sound, such as future high density CD formats.

3.7.4 apt-X 100

apt-X 100 was developed by Audio Processing Technology, a subsidiary of Solid State Logic. It is slightly unusual amongst the systems described here since it is not based first and foremost on auditory masking theory, although it has been shown to perform well when masking theory is applied to the noise analysis. It has its design basis in speech coding, but is designed for high quality music applications. It splits the audio signal into only four subbands, and achieves the additional reduction in data rate by using linear prediction and ADPCM (adaptive differential PCM).

At an audio sampling rate of 32 kHz the resulting data rate is 128 kbit/s per channel, or 192 kbit/s at 48 kHz. Like the Dolby systems, apt-X 100 is not part of the ISO/MPEG draft standard, but has been implemented commercially in a number of recording and broadcasting systems.

3.7.5 IMA ADPCM

The International Multimedia Association (IMA) has been responsible for the difficult job of ensuring a degree of standardisation for multimedia material of various kinds. As described in Table 2.3, it has produced recommendations for a minimum set of sampling rates and quantising methods which are to be supported on any IMA compliant platform. They include the recently specified IMA ADPCM algorithm (based on the Intel/DVI algorithm) which has been designed to be decodable in real time using only the normal CPU facilities of a desktop computer, as opposed to standards like MPEG which require separate digital signal processing. In other words, IMA ADPCM can be played back at reasonably high quality on virtually any modern desktop computer using a software decompression alogrithm.

It allows high quality 16 bit audio to be coded at 4 bits per sample by using a form of differential PCM, in which a predictor attempts to predict the value of the next sample and the algorithm then codes only the difference between the actual sample and the prediction, using 4 bit quantisation and a variable step size which depends on the rate of change of the waveform. On decoding the step size is determined and the sample is converted back to a linear difference value. The same form of prediction is used as during encoding, and the difference value is added to the predicted value to obtain a new sample value which will be similar to that originally encoded.

This algorithm is intended to offer reasonable quality at a data rate one quarter that of the 16 bit rate, whilst requiring only a small overhead in terms of computer CPU time for the decompression process.

Figure 3.13 Copying of
data reduced audio signals.
(a) Decoding and recoding of
the data-reduced signal
results in quality loss. (b)
Copying the data-reduced
signal directly would preserve
quality, provided that source
and destination operated to
the same standard and bit
rate

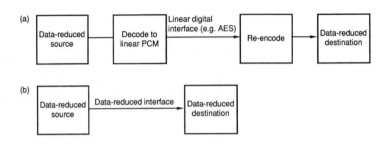

3.8 Quality issues in data reduced audio

It is interesting to consider that what we may witness with data
reduced digital signals in the years to come represents a return
to the problems of the days of analog audio. It will become
important again to assess whether a particular recorder or
broadcast product will give the required sound quality both in
the first generation and after many generations, since this will
depend on the data reduction algorithm and the accuracy of the
psychoacoustic model used to code the signal. Whereas with
linear PCM one may copy a signal digitally without losing
quality (provided that the source and destination operate at the
same resolution), with data reduced signals the subjective
quality may become degraded by copying. Whether or not it
does in practice depends on whether the *coded* signal is copied,
or whether the coded signal is converted back to linear PCM first
(see Figure 3.13). In order to allow copying of the coded signal
a need will arise for standard digital interfaces which handle
data-reduced audio, since current interfaces work only with
linear PCM. Alternatively, in recording and editing systems
based on computer storage media, a simple copy of the
compressed file over a network to its destination would have the
desired result.

Although quality losses can be minimised by ensuring that as
few generations of coding and decoding as possible are involved
in the production and transmission chain, there will inevitably
be stages at which the signal must be converted back to linear
PCM for processing. In the data reduced domain, operations like
filtering and effects are more difficult to realise and require
entirely different digital filters to those used on linear PCM
signals, although the suggestion has been made that considerable savings in DSP power could be gained by operating on
data-reduced signals.[12] The question of whether to convert a
coded signal back to linear PCM is similar to that of converting
a linear PCM signal back to analog, since one has to weigh up
the need against the potential quality reduction. If a certain

process is only available in another domain then there is really no alternative to converting the signal.

The term 'coding margin' is a useful concept, since it refers to how far below the hearing threshold the unwanted side effects of low bit rate coding lie. For example, in an ISO Layer 1 system operating at the relatively high data rate of 192 kbit/s per channel there is plenty of coding margin in the case of nearly all audio signals that one could throw at it. In other words, the additional quantising noise products generated by the coding process are a long way below the masking threshold and in nearly all circumstances will not be heard – therefore there could be said to be a good coding margin. Each successive generation of coding and decoding will gradually raise the unwanted products in level until they pop up above the masking threshold, at which point they will be heard. In a Layer 3 system operating at 64 kbit/s the noise is much closer to the masking threshold and thus sound quality will be noticeably affected after only a small number of generations, and even in the first coded generation may be audible.

If it is envisaged that much conversion will be experienced by a signal between the coded and uncoded domains, then it is advisable to use as high a bit rate as possible to start with. This way the quantising noise will remain under the masking threshold through more generations of coding and decoding. Broadcasting organisations, for example, are recommending that a rate of 192 kbit/s is used for contributions, because they may suffer post-processing and further coding for transmission. The moral of this story is that systems offering large amounts of data reduction, say greater than a factor of six, will be much more likely to show up coding artefacts after copying and post-processing operations than systems offering only modest reductions of say a factor of four.

3.9 Editing and processing of data reduced audio

In most systems described above, audio data are grouped into frames or blocks for coding. Workstations which store sound files in MPEG coded form, for example, are normally restricted in terms of editing accuracy to the MPEG frame resolution, and in some applications this may not be sufficiently accurate. Table 3.2 gives some examples of time resolutions possible with different MPEG sampling rates and layers. It will be seen that Layer 1 gives the best time resolution for editing purposes because of its shorter frames, and that higher sampling frequencies give better resolution than lower ones. Even so the best editing resolution is 8 ms, using Layer 1 at 48 kHz, which is quite a lot

Table 3.2 MPEG time resolutions for editing purposes

MPEG 1 sampling frequencies		
Fs (kHz)	Layer 1	Layer 2
32	12 ms	36 ms
44.1*	~8.71 ms	~26.12 ms
48	8 ms	24 ms
MPEG 2 lower sampling frequencies		
Fs (kHz)	Layer 1	Layer 2
16	24 ms	72 ms
22.05*	~17.42 ms	~52.24 ms
24	16 ms	48 ms

* The frame length varies slightly because the sampling
frequency is not an integer multiple of the compressed bit rate.

coarser than the one sample resolution often possible with linear PCM-based editors. (One sample at 48 kHz is equivalent to a time resolution of 20.8 µs.)

Workstation manufacturers using MPEG for audio storage have spent some time designing processing algorithms which operate on the data reduced audio, so as to avoid the need for repeated stages of decoding and recoding. The French company Digigram, for example, claims to offer the following operations on MPEG audio without decoding to linear PCM:[13]

- file merging with no loss of quality;
- real-time mixing of MPEG sound files;
- time stretching;
- pitch shifting;
- noise reduction;
- equalisation;
- file format conversion.

Audio that has been decoded from a data reduced form to linear PCM and then post-processed should be handled with care. Certain dynamic processing and equalisation could unmask noise that was previously masked. It is worth reiterating that low data reduction factors (high data rates) should be used if post-processing is envisaged.

References

1   Cellier, C. (1993) 'Lossless audio data compression for realtime applications'. Presented at the 95th AES Convention, New York, October 7–10. Preprint 3780.

2   Moore, B.C.J. (1989) *An Introduction to the Psychology of Hearing*, 3rd edition. Academic Press.

3   Gerzon, M. (1992) 'Directional masking coders for multichannel subband audio data compression systems'. Presented at the 92nd AES Convention, Vienna, 24–27 March. Preprint 2TM1.07.

4   Fletcher, H. (1940) 'Auditory patterns'. *Reviews of Modern Physics*, **12**, pp. 47–65.

5   Zwicker, E. (1961) 'Subdivision of the audible frequency range into critical bands'. *Journal of the Acoustical Society of America*, **33**, 2, p. 248.

6   Smyth, M. and Smyth. S (1991) 'APT-X 100: a low-delay, low bit-rate, sub-band ADPCM audio coder for broadcasting'. In Proceedings of the AES 10th International Conference, London, 7–9 September, pp. 41–56.

7   International Standards Organisation (1993) *ISO/IEC 11172: Information Technology – Coding of moving pictures and associated audio for digital storage media at up to about 1.5 Mbit/s.*

8   Stoll, G., Nielsen, S. and van de Kerkhof, L. (1993) 'Generic architecture of the ISO/MPEG layer 1 and 2: compatible developments to improve the quality and addition of new features'. Presented at the 95th AES Convention, New York, 7–10 October. Preprint 3697

9   International Standards Organisation (1994) *ISO/IEC DIS 13818-3 – Generic coding of moving pictures and associated audio*. March.

10   Stoll, G. *et al* .(1993) 'Extension of ISO/MPEG-Audio Layer 2 to multichannel coding'. Presented at the 94th AES Convention, Berlin, 16–19 March. Preprint 3550.

11   Grill, B. *et al* .(1994) 'Improved MPEG-2 Audio multichannel encoding'. Presented at the 96th AES Convention, Amsterdam, 26 Feb–1 Mar. Preprint 3865.

12   Brandenburg, K. and Herre, J. (1992) 'Digital audio compression for professional applications'. Presented at the 92nd AES Convention, 24–27 March, Vienna.

13   Digigram (1995) T*he PCX application guide*. Digigram, Parc de Pré Milliet, 38330 Montbonnot, France.

4 Data storage media

The purpose of this chapter is to describe the principles, limitations and applications of storage media used for digital audio in computer workstations. The media described are not exclusive to the field of audio and are widely encountered in general purpose storage applications. In most cases the same media that are used for general purpose applications in computers can be used for storing audio and even video without modification, although certain criteria must be observed concerning their specifications if operation is to be satisfactory. The ability to use computer mass storage media for audio and video is advantageous because it allows these industries to benefit from the economies of scale evident in the computer industry. Audio and video alone could not warrant the research, development and mass production engineering required to make high performance storage media at low cost. There will continue to be a decline in the use of dedicated audio recording formats in favour of general purpose mass storage media, if only because of the simple economics of the matter.

There can be no doubt that there will continue to be improvements in the design of storage media, and that capacity and speed will continue to increase whilst prices fall. There is no sign of a slowing down in the rate of development at the moment. The devices described here are likely to remain popular for some years to come, and in any case the fundamental principles involved are unlikely to change radically. Examples of specifications should only be taken as representative of today's equipment.

4.1 Storage requirements of digital audio and video

Before looking at specific storage options it would be as well to investigate the storage requirements for multimedia data such as audio and video. Certain storage devices can then be ignored because they do not fulfil the requirements.

There are two main roles for storage media here. One is the primary role of real-time recording and replay, and the other is the secondary role of backup storage. The requirements differ somewhat, although it is possible to use similar media for both purposes. Real-time recording and replay needs storage devices capable of sustaining data transfer for a number of audio channels, so that the channels can record or replay for long periods without breaks, be edited and post-processed, and with quick access to stored files. Backup can take place in non-real time, does not need such fast access to files, and does not need to support editing and other post processing operations. Backup may also need a large capacity and it would be advantageous if it were cheaper than primary storage, and be based on removable media. It follows that certain devices are suitable for backup that may not be suitable for primary storage.

Table 4.1 shows the data rates required to support a single channel of digital audio at various resolutions, and Table 4.2 shows the requirements for data-reduced forms at different bit rates. Media to be used as primary storage would need to be able to sustain data transfer at a number of times these rates to be useful for multimedia workstations. The tables also show the number of megabytes of storage required per minute of audio, showing that the capacity needed for audio purposes is considerably greater than that required for text or simple graphics

Table 4.1 Data rates and capacities for linear PCM

Sampling rate kHz	Resolution bits	Bit rate kbit/s	Capacity/min. Mbytes/min	Capacity/hour Mbytes/hour
96	16	1536	11.0	659
88.1	16	1410	10.1	605
48	20	960	6.9	412
48	16	768	5.5	330
44.1	16	706	5.0	303
44.1	8	353	2.5	151
32	16	512	3.7	220
22.05	8	176	1.3	76
11	8	88	0.6	38

Table 4.2 Data rates and capacities for data-reduced audio

Bit rate kbit/s	Capacity/min. Mbytes/min	Capacity/hour Mbytes/hour
64	0.5	27
96	0.7	41
128	0.9	55
196	1.4	84
256	1.8	110
384	2.7	165

applications. Storage requirements increase pro rata with the number of audio channels to be handled. The rates and capacities required for video can be even higher, and are discussed in more detail in Chapter 7.

Storage systems may use removable media but many have fixed media. It is advantageous to have removable media for audio and video purposes because it allows different jobs to be kept on different media and exchanged at will, but unfortunately the highest performance is still obtainable from storage systems with fixed media. Although the performance of removable media drives is improving all the time, fixed media drives have so far retained their advantage. Systems involving a small number of audio channels or using data reduction may be able to take advantage of removable media as primary storage, but in most current systems removable media are normally used as secondary storage.

It perhaps goes without saying that any storage system used for audio and video should be as reliable and robust as possible. It is also likely to need to be a fairly 'heavy duty' system because the demands of audio and video recording are quite heavy and will require the storage device to be in an almost constant state of activity, as opposed to the more gentle task of, say, word processing, where the storage device is idling for long periods.

4.2 Disk drives in general

Disk drives are probably the most common form of mass storage. They have the advantage of being random-access systems – in other words any data can be accessed at random and with only a short delay. This may be contrasted with tape drives which only allow linear access – that is by winding through the tape until the desired data is reached, resulting in a considerable delay. Disk drives come in all shapes and sizes

Figure 4.1 The general mechanical structure of a disk drive

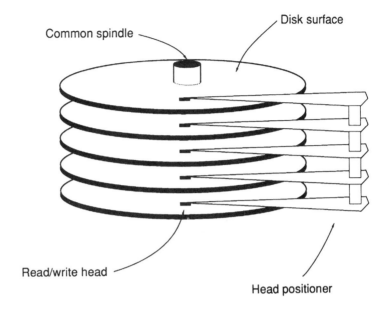

Common spindle

Disk surface

Read/write head

Head positioner

from the commonly encountered floppy disk at the bottom end to high performance hard drives at the top end. The means by which data are stored is usually either magnetic or optical, but some use a combination of the two, as described below. There exist both removable and fixed media disk drives, but in almost all cases the fixed media drives have a higher performance than removable media drives. This is because the design tolerances can be made much finer when the drive does not have to cope with removable media, allowing higher data storage densities to be achieved. Although removable disk media can appear to be expensive compared with tape media, the cost must be weighed against the benefits of random access and the possibility that some removable disks can be used for primary storage whereas a tape cannot.

The general structure of a disk drive is shown in Figure 4.1. It consists of a motor connected to a drive mechanism which causes one or more disk surfaces to rotate at anything from a few hundred to many thousands of revolutions per minute. This rotation may either remain constant or may stop and start, and it may either be at a constant rate or a variable rate, depending on the drive. One or more heads are mounted on a positioning mechanism which can move the head across the surface of the disk to access particular points, under the control of hardware and software called a disk controller. The heads read data from and write data to the disk surface by whatever means the drive employs. It should be noted that certain disk types are read-only,

Figure 4.2 Disk formatting divides the storage area into tracks and sectors

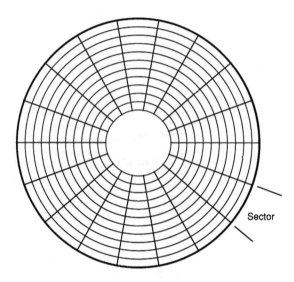

Sector

some are write-once–read-many (WORM), and some are fully erasable and rewritable.

The disk surface is normally divided up into tracks and sectors, not physically but by means of 'soft' formatting (see Figure 4.2). Formatting places logical markers which indicate block boundaries, amongst other processes. On most hard disks the tracks are arranged as a series of concentric rings, but with some optical disks there is a continuous spiral track.

4.3 Disk drive specifications

Disk drive performance is characterised by certain specifications which are often quoted in promotional literature. These are the subject of a certain amount of misunderstanding and manufacturers often play games with these figures to make their drives seem better than they are. As with all specifications it is important to compare like with like, and to know how a certain parameter has been measured. The most important parameters are:

- access time;
- instantaneous transfer rate;
- sustained transfer rate;
- storage capacity (formatted).

These are not the only factors that affect the performance or desirability of a drive, but they are a ready means of comparing two apparently similar drives.

Figure 4.3 The delays involved in accessing a block of data stored on a disk

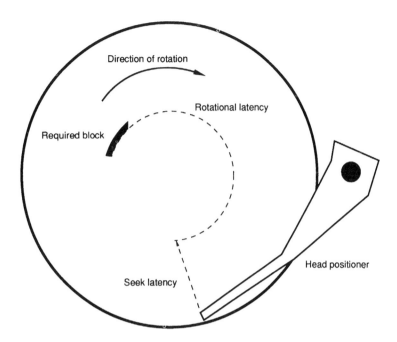

Access time, normally quoted in milliseconds, is the time taken for a block of data to be accessed. It may be specified in a number of ways, since clearly the actual access time depends on where the head is when a block is requested. Figure 4.3 shows that true access time is made up of *seek latency* and *rotational latency*. The seek latency is dependent on the speed of the positioner and the rotational latency is dependent on how fast the disk rotates. Access time may often be just seek latency, and may be quoted as 'track-to-track', which is the fastest; 'average', which is a reliable guide to general performance; or 'one-third full sweep', which is the time taken for the head to traverse one third of the active disk radius.

Instantaneous transfer rate is the fastest speed at which data can be read from the disk surface once the head has arrived at its correct location. Normally quoted in megabits per second it gives a guide to the peak performance of the drive. Sustained transfer rate is a more useful guide to real performance, though, because it gives a guide to the long-term data rate which might be expected from the disk, sustained over many blocks. This parameter, though, is affected considerably in real multimedia systems by the fragmentation of the drive and by the number of channels it has to service, as discussed in section 5.2.3.

Formatted storage capacity is the number of megabytes of capacity available for user data after the disk has been formatted (see section 4.11). It is often considerably smaller than the unformatted capacity of the disk (which is not a very useful figure to know). The formatted capacity is thus all available for the storage of audio data if necessary, with no necessity to add an overhead for error correction, as described in Chapter 2.

4.4 Magnetic disk drives

4.4.1 Winchester hard disk drives

The Winchester magnetic disk drive has been used in personal computers for quite some time now. It provides space for the storage of a large amount of data in a relatively small space, is reliable, fast and reasonably economical. The Winchester drive is a sealed unit, and the physical disks inside it cannot be removed to make way for others. The recording process is magnetic, whereby data is stored in the form of flux reversals in the surface layer of the disks. The drive is a combination of physical disk surfaces on which data is stored, electromagnetic heads which read and write data, a positioner to move the heads to the right place, a motor which rotates the surfaces, a servo mechanism which controls the moving parts, and a controller which looks after the data flow to and from the surfaces, and interfaces to the rest of the computer system. A cut-away diagram of a drive is shown in Figure 4.4.

The Winchester drive is sealed (except sometimes for a small pressure-relief vent) in order to prevent the surfaces of the disks from becoming contaminated. The lack of contamination and the

Figure 4.4 Cut-away drawing of a typical Winchester drive. (Courtesy of MacUser)

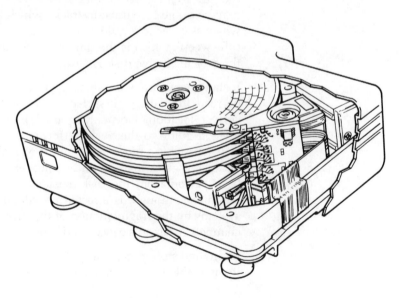

fact that the disks will never be removed means that fine tolerances can be used in manufacture, allowing a larger amount of data to be stored in a smaller space than is possible with removable magnetic disks. It also results in a very low error rate. More than one disk resides inside a Winchester drive, and it is common for both sides of each disk to be used. These disks are rigid, not floppy, and all rotate on a common spindle. Each surface has its own 'pickup', or read/write head, which can be moved across the disk surface to access data stored in different places. The head positioner moves all the heads at the same time, rather than independently. The heads do not touch the surface of the disks during operation, they fly just a small distance above the surface, lifted by the aerodynamic effect of the air which is dragged around above the disk surface due to friction. A small area of the disk surface is set aside for the heads to land on when the power is turned off, and this area does not contain data.

Data is stored in tracks divided up into sectors. Each sector is separated by a small gap and preceded by an address mark which uniquely identifies the sector's location, and a preamble to synchronise the reading of data. The term *cylinder* relates to all the tracks which reside physically in line with each other in the vertical plane through the different surfaces (see Figure 4.5). A sector typically contains 512 bytes.

Figure 4.5 Winchester drive tracks on different surfaces form concentric cylinders

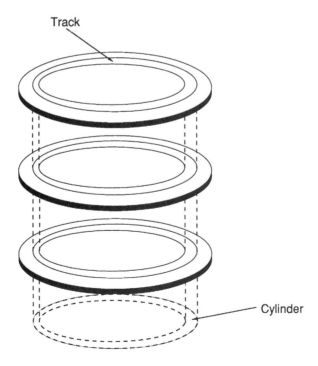

The specifications of hard disks are improving all the time, and thus any figures quoted are unlikely to be realistic within a short time. None the less it may be interesting to note that, at the time of writing, drives are available with storage capacities of a number of gigabytes (1 Gbyte = 1000 Mbytes), access times of only a few milliseconds (<10 ms is possible), and transfer rates of many tens of megabits (or even megabytes) per second. This makes them eminently suitable for most professional audio applications, and many multimedia applications involving compressed video. (Uncompressed video involves data rates too high for the majority of desktop systems.) One can store many hours of monophonic audio on a hard disk, and they are capable of handling a number of simultaneous channels of recording and replay.

The disk is of the 'write-many-times' format which means that old data may be overwritten many times in order to reuse the storage space. Backup is normally performed by copying data from the drive to a removable medium such as a tape cartridge or optical disk. Although the disk surfaces of a Winchester are not removable, drives exist which may be interchanged in their entirety. Such drives are known as removable drives (not removable disks), and they are usually mounted in a cartridge with a handle so that they may be 'unplugged' from a docking frame of some sort. Figure 4.6 shows a photograph of such a system. Whilst this is quite a useful feature, it is relatively expensive to interchange complete drives in this way, although it may be considered worth the advantage of being able to take a complete session's primary storage from one system and insert it into another.

Winchester disk drives have now been developed to the point where high capacities are available in very modestly sized packages, and they may be installed within desktop or laptop machines. A quiet drive is important for audio operations, especially if the drive is to be installed in the same room as the operator, and drives vary considerably as to their noisiness.

4.4.2 AV drives

A range of hard disk drives has been introduced specifically to serve the industry demand for storage media suitable for audio and video applications. There are a number of aspects of these drives which have been optimised specifically for the increased data throughput requirements.

Modern high capacity Winchester drives often require recalibration at regular intervals to compensate for thermal changes which may affect the accuracy of head positioning, as well as

Figure 4.6 A typical removable disk drive system allowing multiple drives to be inserted or removed from the chassis at will. (a) Frame housing multiple removable drives. (b) Removable storage cannisters. (Courtesy of Tyrell Corporation)

(a)

(b)

other 'housekeeping' activities such as head demagnetisation. The recalibration process is instituted automatically by the disk controller, often no matter what else is going on at the time, and it may take over 100 ms. This can interrupt the flow of data to or from the drive. Whilst this is relatively unimportant when loading a text file, it becomes significant when real-time recording or replay of audio and video is involved. The calibration process reduces the overall data flow temporarily which may affect smooth replay or recording. Audio recording and replay is normally memory buffered (see section 5.2.2), which has the effect of ironing out short-term irregularities in data flow, but the built-in video software of many multimedia computers (such as Apple's QuickTime) is often slowed down by such events. By using more advanced recalibration procedures, which intelligently monitor requests for data transfer, AV drives ensure that recalibration does not coincide with data transfers.

Error recovery on ordinary disk drives may involve the re-reading of erroneous sectors, and this also has the effect of causing a short delay in data transfer. AV drives may have improved error handling and improved mechanics which can stand up to the wear and tear of data intensive tasks. Cache memory in the disk controller allows the transfer of data to be optimised. The data blocks required for a particular file transfer may not be stored contiguously on the disk surfaces, requiring some head movements to reassemble the file. Optimum data transfer might require that blocks be read or written in the 'wrong' order because of the advantage that would be gained by reading or writing physically close blocks in sequence, and the temporary RAM cache is used as a place to put these blocks so that they can subsequently be handled in correct sequence.

There is usually a small price premium attached to AV drives because of the above features.

4.4.3 RAID arrays

Hard disk drives can be combined in various ways to improve either data integrity or data throughput. RAID stands for

Table 4.3 RAID levels

RAID level	Features
0	Data blocks split alternately between a pair of disks, but no redundancy so actually less reliable than a single disk. Transfer rate is higher than a single disk. Can improve access times by intelligent controller positioning of heads so that next block is ready more quickly
1	Offers disk mirroring. Data from one disk is automatically duplicated on another. A form of real-time backup
2	Uses bit interleaving to spread the bits of each data word across the disks, so that, say, eight disks each hold one bit of each word, with additional disks carrying error protection data. Non-synchronous head positioning. Slow to read data, and designed for mainframe computers
3	Similar to level 2, but synchronises heads on all drives, and ensures that only one drive is used for error protection data. Allows high speed data transfer, because of multiple disks in parallel. Cannot perform simultaneous read and write operations
4	Writes whole blocks sequentially to each drive in turn, using one dedicated error protection drive. Allows multiple read operations but only single write operations
5	As level 4 but splits error protection between drives, avoiding the need for a dedicated check drive. Allows multiple simultaneous reads and writes
6	As level 5 but Incorporates RAM caches for higher performance

Redundant Array of Inexpensive Disks, and is a means of linking ordinary disk drives under one controller so that they form an array of data storage space, as shown in Figure 4.7. A RAID array can be treated as a single volume by a host computer. There are a number of levels of RAID array, each of which is designed for a slightly different purpose, as summarised in Table 4.3.

One of the main reasons for using a RAID array would be to improve the reliability of data storage. At certain RAID levels

Figure 4.7 Some examples of RAID array configurations. (a) Level 0; (b) Level 1; (c) Level 3

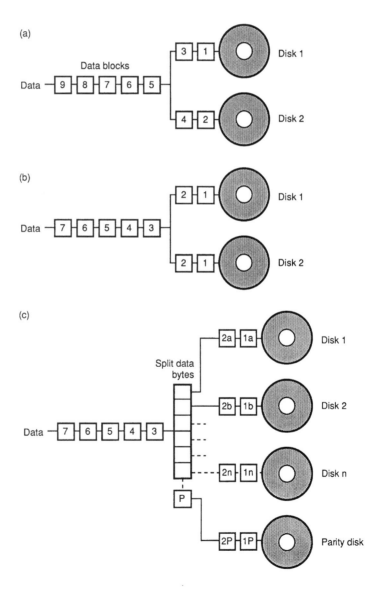

the data is spread across all of the drives involved, with a final drive used to store error protection information (the check drive). The aim is to prevent you losing your data if one of the drives fails, because it can be reconstructed from the remaining data. 'Mirroring' is also an option which allows the data on one disk to be perfectly duplicated on another, again for improving data security. By spreading data across drives it is also possible to speed up read and write operations.

4.4.4 Removable magnetic media

Floppy disks are unsuitable for AV applications because of just about every aspect of their specification. They are too small and too slow. Higher capacity removable magnetic media have existed for some time, though, with speeds approaching that of slower hard disks. The most well known are Bernoulli and Syquest drives, which are constructed rather like large floppy disks, in a rigid cartridge. These have tended to offer capacities of under 100 Mbytes, which makes them only marginally useful for AV applications requiring short storage times.

Recent advances in the magnetic recording field have resulted in removable media offering much higher capacities and transfer speeds. One example, the Iomega *Jaz* drive, boasts performance which suggests that for the first time a removable magnetic disk may be suitable for primary storage in AV systems. The specification includes a sustained transfer rate of up to 6.73 Mbytes per second, 12 ms seek time and 17.5 ms average overall access time. Disks will be in the form of a dual platter 3.5 inch cartridge, with capacities of either 540 Mbytes or 1 Gbyte.

4.5 Optical disks

4.5.1 General

There are a number of families of optical disk drive, which have differing operational and technical characteristics, although they share the universal benefit of removable media. They are all written and read using a laser, which is a highly focused beam of coherent light, although the method by which the data is actually stored varies from type to type.

The question of whether or not disks may be interchanged between optical drives requires some discussion, since the method of formatting and the precise method of data pickup may differ. The most obvious difference lies in the erasable or non-erasable nature of the disks. WORM disks may only be written once by the user, after which the recording is permanent. Most Compact Disc formats are not even Write-Once (at least from the user's point of view), since they are stamped at a factory and act as a read-only medium, although the CD-R (recordable CD) is now available, being a CD-compatible disk which may be recorded once but not erased. Optical disks (except CDs) are normally enclosed in a plastic cartridge which protects the disk from damage, dust and fingerprints.

The magneto-optical (M-O) disk is similar in appearance to the WORM, but uses technology which allows data to be erased and the surface rerecorded. The speed of magneto-optical drives is beginning to approach the speed of slow Winchester drives, which opens up possibilities for using them as a direct alternative to the Winchester in desktop multimedia systems. One of the major hurdles which had to be overcome in the design of such optical drives was that of making the access time suitably fast, since an optical pickup head is much more massive than the head positioner in a magnetic drive (it weighs around 100 g as opposed to around 10 g). Techniques are being developed to rectify this situation, since it is the primary limiting factor in the onward advance of optical storage.

Drives are now available which will read and write both WORM disks and M-O disks, provided that they conform to the ISO standard.

4.5.2 CAV and CLV modes

There are two modes of rotation used variously in optical disk drives. CAV (Constant Angular Velocity) recording and CLV (Constant Linear Velocity) recording. In CLV recording the rotational speed of the disk changes depending on the position of the pickup, in order to keep a constant length of track passing under the head per second. In CAV recording the rotational speed of the disk remains constant. CAV disks normally have sectors of a fixed angle of arc, holding a fixed amount of data, and therefore the data is more densely packed in sectors towards the centre of the disk (see Figure 4.8). CLV recording allows more data sectors to be stored towards the edges of the disk than at the centre, and may therefore allow more efficient use to be made of the space available, but CLV disks require servo operation to change the disk speed when the pickup head is moved, making them slower to access data.

Some drives use a mode known as zoned-CAV (Z-CAV) to pack more data into the outer tracks of a disk, wherein the disk rotates

Figure 4.8 (a) Sectors on a CAV disk are of equal angle of arc. (b) On a Z-CAV disk the sector angle is not constant and more sectors are recorded at the outer edges of the disk than at the centre

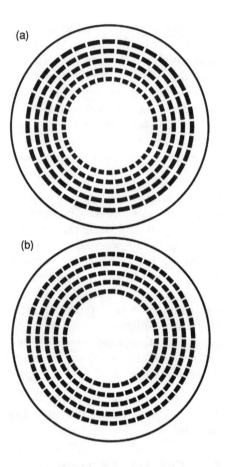

(a)

(b)

at one of a number of fixed speeds depending on which 'zone' the pickup is in. This is really a halfway house between CAV and CLV recording, and does not compromise access time so much.

Compact Discs use CLV recording, as do certain WORM drives, but most computer optical disk drives use a form of CAV or Z-CAV recording.

4.5.3 The WORM drive

The WORM writing process involves the one-time modification of the disk's recording surface, such that the reflectivity of the surface is altered. The disk is made up of a sandwich of protective layer, reflective layer, recording layer and polycarbonate substrate (see Figure 4.9). In the Sony WORM cartridge, the recording layer is itself a sandwich of bismuth–tellurium and antimony selenide, which is claimed to offer a storage life of over

Figure 4.9 Cross-section through a typical WORM disk

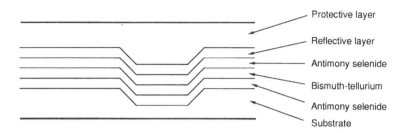

Protective layer
Reflective layer
Antimony selenide
Bismuth-tellurium
Antimony selenide
Substrate

Figure 4.10 A WORM disk and drive. (Courtesy of Panasonic UK)

100 years, an important consideration since WORMs are ideal as an archival medium. In order to write data to the disk a modulated laser heats small areas of the recording layer, forming an alloy which has a different reflectivity pattern to the unheated areas. The surface of a WORM disk is 'pre-grooved' and formatted, to provide a guide track for the drive to follow

when writing data. On replay, lower-powered laser light is reflected back from the surface with a varying intensity which depends on the presence or lack of recorded areas. A WORM drive and disk are shown in Figure 4.10.

A number of different formats for WORMs exist, and not all disks are compatible between different manufacturers' systems. One manufacturer, for example, offered two drives which differed considerably in performance and capacity. Each required a differently-formatted optical disk. One drive used disks which held 200 Mbytes per side, using conventional sectors of 512 bytes and CAV recording, whereas the other offered greater storage capacity by using Z-CAV recording. The disks for the second drive had 1024 byte sectors and held 470 Mbytes per side. The rotation speeds were also different, being 875 rpm in the former and 1200 rpm in the latter. The performance of the latter drive was much more spectacular than the former, offering an instantaneous data transfer rate at the optical head of 5.55 Mbit/s (average) compared with 2.5 Mbit/s. From this it may be seen that the world of WORMs is not a straightforward one.

A WORM disk may be written to until its capacity is filled, the directory of contents being added to as the number of files increases. Since the WORM disk may not be erased, data is very secure, and sound recordings committed to WORM would be free from the danger of accidental erasure. Files are stored sequentially, that is blocks of data are recorded one after the other in numerical sequence, with new files being recorded immediately after the end of the last file on the disk. There is no danger of the disk becoming fragmented since old data is never erased. These factors add up to making the WORM quite suitable for the storage of mono or stereo sound files, but perhaps less suitable for multi-channel operation. Whether or not real-time editing is possible from a WORM depends on the performance of the drive: an average transfer rate of at least 4 Mbit/s is sensible as a minimum for flexible stereo operation.

4.5.4 The magneto-optical (M-O) drive

M-O drives use optical disks which may be erased and rerecorded. In order to write data, the laser is used at a higher power to that used in the reading process, again to heat spots in the recording layer which is made up of rare earth elements (typically gadolinium and terbium). A biasing magnet is used to create a weak magnetic field in the vicinity of the heated spot on the disk, whose recording layer only takes on this prevailing magnetic polarisation when it is hot, whereas under normal

Figure 4.11 The magneto-optical disk is recorded by exposing small areas of the recording layer to high-power laser light, whereupon they take on the magnetic polarity provided by the polarising magnet. On replay, the magnetic polarisation affects the polarisation of reflected laser light

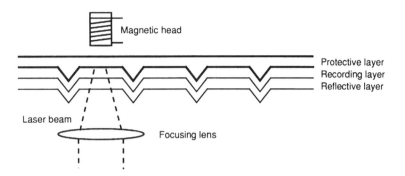

conditions the recording layer cannot be magnetised (see Figure 4.11). When the spot cools it retains this magnetisation. Although the data is recorded by a combination of optical heating and magnetisation, it is read by an entirely optical means which relies upon the fact that laser light reflected from the disk will be polarised depending on the magnetic polarisation of the recording layer. This is known as the Kerr effect, and the change in optical polarisation angle may be as small as a few degrees depending on the material concerned. The reflected light passes through a polarisation analyser, resulting in changes in intensity of the light falling on a photodetector. The M-O disk is pre-grooved and sectored like the WORM drive.

The disk may be erased by heating the relevant area as it passes under the laser head, whilst biasing it with the opposite magnetic field direction to that used for recording, after which it may be rewritten by the same process as before. It is this erase cycle, required before rewriting, which makes M-O drives slower to write data than to read it, whereas Winchester drives erase old data simply by overwriting it with new data. In most drives, a permanent magnet is used to provide the biasing field which magnetises the spots on the disk when they are heated, the magnet being physically rotated between erase and write cycles to change the direction of the field (this can be achieved in a few milliseconds). Spots on the surface of the disk are magnetised or not by turning the laser on and off – only spots which are heated will have their magnetisation changed. Hitachi suggested a solution in late 1987 which appeared to allow rewriting without the primary erase revolution of the disk. It involved the use of an electromagnet rather than a permanent magnet, on the opposite side of the disk to the laser. In order to rewrite, the laser heats the relevant areas of the disk and the magnetic field direction is switched electrically between −300 and +300 oersteds. It was said that this meant that recording could only take place on one side of the disk, and that oxidation of the disk surface would occur more quickly, shortening its life. 'Background erasure' has also been attempted by some designers trying to incorporate M-O drives into audio systems, which is a means of erasing old files 'in the background' whilst the drive is not being used for audio transfer, thus freeing up space for new files which may then be written without requiring the erase pass.

An ISO standard was established for M-O drives, to which most of the major manufacturers adhere. This allows for two different sector sizes (512 bytes and 1024 bytes), giving 297 and 325 Mbytes per side of storage capacity respectively on a 5.25 inch disk (594 or 650 Mbytes in total) using CAV recording. There is a higher density version offering 1.2 or 1.3 Gbytes of storage, and a 2.6 Gbyte version is imminent at the time of writing. In addition to the 5.25 inch M-O cartdridge there also exist 3.5 inch cartridges capable of storing 128 Mbytes or 256 Mbytes (high density). This is quite a small capacity for uncompressed digital audio, but the disks are convenient and have quite fast access times. The cost of M-O disks has also fallen to quite a reasonable level.

The performance of M-O drives is reasonably good, but still not as good as Winchester disks, with slower access times and transfer rates. Some manufacturers have deviated from the ISO specification in order to achieve faster access and greater capacity, using Z-CAV techniques, and these drives are usually capable of operating in both standard and non-standard modes. M-O drives may be used with care as the primary storage medium for some digital audio systems, but only for the recording and replay of a limited number of channels. This will depend largely on the recommendation of the manufacturer.

4.5.5 Interchange of M-O disks

M-O disks which conform to the ISO standard should be able to be read by other drives which conform to the same standard. M-O drives do not all rotate at the same speed, although this should not affect interchange, since the data structure on the disks should be the same, only the transfer rate will be better on faster drives. Whether or not this interchangeability applies to audio systems using M-O drives is rather another question, because the way in which sound files are stored on the disks will differ between each system. There are a number of levels at which compatibility must exist before total interchangeability can be achieved between audio systems. This is discussed in greater detail in section 4.10.

4.5.6 Phase-change drives

In phase-change recording data is written by a high-power laser, in much the same way as with a WORM, changing spots from a non-crystalline (amorphous) state to a crystalline state. In the crystalline state the reflectivity is increased considerably over that of the amorphous state. Data is again read back by a lower-power laser which detects changes in reflectivity. So the process is very WORM-like, but it is claimed that by careful selection of the recording material and laser beam control the process may be made reversible (and thus data may be overwritten). The only apparent drawback is the number of re-write cycles allowed (cycles of erasure and re-recording), which is quoted by Panasonic as being an order of ten lower than that of the M-O disk. Phase-change disks are not compatible with M-O drives, and vice versa.

4.5.7 Compact Discs and drives in general

Compact Discs (CDs) are familiar to most people as a consumer read-only optical disk for audio storage. They are also used in other ways, and the number of different CD formats is now quite large, resulting in a plethora of disk and drive types, not all of which are compatible with each other. Table 4.4 is an attempt at a summary of current CD formats and their applications. Most of the formats are described in coloured 'books' and many have been defined principally by Philips and Sony (sometimes in conjunction with others). Most CDs are read only and are used for distributing data cheaply, having a maximum storage capacity of about 650 Mbytes. They have wide applicability in multimedia systems for desktop computers, and newer machines now incorporate CD drives as a standard feature. A high density CD is in development, which is likely to increase capacity to around 4.7 Gbytes, and is intended for high quality MPEG-2 video and high quality multichannel audio applications as well as advanced multimedia purposes.

The CD-ROM standard divides the CD into a structure with 2048 byte sectors, addressable by a computer, adds an extra layer of error protection, and makes it useful for general purpose data storage including the distribution of sound and video in the form of computer data files. There is a standard filing structure for CD-ROM known as ISO 9660 or High Sierra, which is used when wanting to ensure that disks may be read across a wide range of platforms, although CD-ROMs can also be formatted in non-ISO modes for use on proprietary platforms. It is possible to find disks with mixed modes, containing sections in CD-ROM

Table 4.4 CD formats

Common name	Book	Applications
CD-DA or CD-Audio	Red	Audio only @ 44.1 kHz, 16 bits
CD+G	Red	As CD-DA but with still graphics stored in the R–W subcodes of the CD
CD-ROM (Mode 1)	Yellow	General purpose data storage applications in either ISO 9660 or non-ISO structures
CD-ROM (Mode 2) or CD-ROM/XA	Yellow (suppl.)	Extra multimedia features for CD-ROM
CD Extra	Blue (part)	CD-DA with CD-ROM data in the spare capacity of an audio album. A means to enhance CD-DA with interactive material. CD-DA data is indexed in a standard Red Book TOC, and comes first on the disc. Computer data follows, but indexed only to Orange Book standard (multisession)
Photo CD	Yellow+/Bridge	High quality still photos. Also playable on CD-ROM/XA Form 1 drives, and CD-I players
Video CD	White/Bridge	70 minutes of high quality MPEG-1 compressed video. Playable on CD-ROM/ XA Form 2 drives and CD-I Digital Video players
CD-I	Green	Interactive consumer applications
CD-I Digital Video	Green/Bridge	Gives CD-I players ability to replay real-time MPEG-1 compressed Video CDs (Full Motion Video, or FMV)
CD-I Ready	Red/Green	CD-I with additional audio-only tracks
CD-R	Orange	Recordable CD. Either CD-WO (write-once) or CD-MO (erasable). Can be used to record CDs in any format with appropriate tools. Not so much a format as a medium
'Rainbow CD'	Mixed	Data from various 'books' mixed on a single CD
Super density CD	–	Recently combined Toshiba/WEA and Philips/Sony proposal for a new high density disk to carry MPEG-2 video and multichannel compressed audio, amongst other things. Name not finalised at time of writing. May also carry high quality uncompressed audio

Note: the Bridge Disc specifications were defined by Philips and Sony to allow Video CDs and Photo CDs to be replayed on a wide range of player types. These standards effectively bridge the CD-ROM/XA and CD-I standards, blurring the distinction between them. The White Book Video CD standard, for example, is an element of the Bridge Disc specification, and is designed to be playable on Mode 2, Form 2 CD-ROM/XA drives or Digital Video-capable CD-I players.

format and sections in CD-Audio format. The recently developed CD Extra is one such example.[1]

The CD drives found in computer systems are usually capable of replaying at least CD-Audio (Red Book) and CD-ROM (Yellow Book). Many are capable of handling CD-ROM/XA and PhotoCD, with multisession capability (they can read disks which were recorded in more than one session), and some have the XA Mode 2/Form 2 needed to replay MPEG-1 Video CDs. The most flexible drives will have the capacity to read both Red and Orange book tables of contents (TOCs), so as to allow them to read data from disks indexed in different ways. The Orange Book TOC is a temporary TOC that allows disks to be recorded in more than one session, as described below. Until recently, access times and transfer rates from CD were quite slow owing to the low rotational speed of between 200 and 500 rpm and use of CLV recording. An important development, though, has been the introduction of higher speed CD drives which improve these parameters quite considerably and make the CD more useful for the replay of multimedia material.

Increasingly, drives are equipped with a facility which allows digital audio from CD-Audio disks to be transferred directly over a SCSI interface to other mass storage media, and stored as a sound file that can subsequently be used in editing operations and other multimedia sound applications. This normally requires software installed on the host computer capable of carrying out this task, and a CD drive with this feature.

4.5.8 CD-R

CD-R is the recordable CD, and may be used for recording CD-Audio format or other CD formats using a suitable CD Recorder and software. CD Recorders are now available as drives which can be attached directly to the SCSI port of desktop computers. Standard audio CDs (CD-DA) conform to the Red Book standard published by Philips, whereas CD-R is described in the Orange Book. The Orange Book contains information on the additional features of CD-R, such as the area in the centre of the disk where data specific to CD-R recordings is stored. Audio CDs recorded to the Orange Book standard can be 'fixed' to give them a standard Red Book table of contents (TOC), allowing them to be replayed on any conventional CD player. Once fixed into this form, the CD-R may not subsequently be added to or changed, but prior to this there is a certain amount of flexibility, as discussed below.

The Orange Book allows for two potential formats: one is basically a WORM format (for disks which can only be recorded once), and the other is an erasable format based on M-O (magneto-optical) disks. Only the WORM disks can currently be replayed on today's conventional CD players, since the principles of M-O replay are slightly different to those involved in a standard pickup, but there is no reason why a dual-standard player should not be introduced in the future, replaying M-O and WORM disks. The current generation of low-cost CD-R machines, though, only work with WORM disks.

Figure 4.12 shows the cross-section through a typical blank CDR WORM disk, and it will be seen that the disk consists of a pre-formed 'groove' in the so-called recording layer. The recording layer consists of a green semi-transparent material, behind

Figure 4.12 Cross section through a CD-R WORM disk

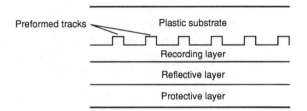

which is a gold reflective layer. During recording, the laser heats the recording layer to around 250 °C, a process which causes it to melt, forming a pit similar to that found on a conventional CD. On replay, the laser pickup, operated at a lower power than for recording, experiences a lower level of reflected light in the presence of a pit than it does in the absence of a pit, in exactly the same manner as for a pre-recorded CD.

Since the CD-R servo system must have some reference to lock to when recording a blank (information which is normally extracted from the data stream in pre-recorded CDs), the pre-grooved track is also modulated to a very slight extent by a sine wave at half the audio sampling frequency (22.05 kHz). The amount of sinusoidal deviation of the groove is only ±0.03 μm, whilst the track width remains at the standard 0.6 μm and the spacing between the tracks remains at 1.6 μm. Superimposed on the pre-groove modulation is information about the running time of the CD, in the form of a ±1 kHz frequency modulation of the 22.05 kHz sine wave.

An Orange Book CD-R does not have to be recorded all at once. It can be removed from the machine and added to at a later date, appending the new track to the end of the last recording, giving it the next-numbered track ID. In order to make this possible the disk contains an additional recording area inside the starting point of a conventional CD (normal CD's begin with a TOC in the centre of the disk and play from the inside out), which is divided up into two parts (see Figure 4.13). The Program

Figure 4.13 Division of recording area on the CD-R, showing space for program calibration area (PCA) and temporary program memory area (PMA)

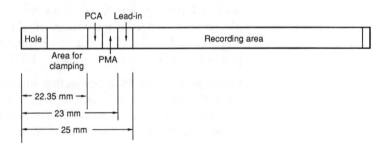

Calibration Area (PCA) is used for optimising laser power by making a number of short test recordings when a new disk is first inserted. On subsequent occasions this calibration is not required since a message is stored on the disk to indicate the appropriate laser power. The Program Memory Area (PMA) is used to store a temporary TOC whilst the disk is yet 'unfixed', and this TOC is updated every time a new track is recorded. Here is also stored 'skip' information, which allows certain tracks to be skipped on replay if they have been messed up (although this may only work when the disk is replayed either on the CD-R machine itself, or on a very recent CD player which recognises skip IDs).

The lead-in area of the CD-R, where a normal CD would start to read its TOC, is left blank until such time as the user decides that the disk is completed. On 'fixing' the disk the machine records a Red Book TOC, after which no further recording is allowed. Thus a CD-R blank still has as much space for audio recording as a conventional CD. The first blanks for these machines run to 63 minutes, but 74 minute disks are also available, running at the slightly slower linear velocity of 1.2 m/s (which one company suggests might make it slightly more prone to replay errors than a 63 minute disk).

A number of recording modes are possible on most CD-R drives. 'Disk-at-once' is the most basic, in which all of the information is written at one time together with a Red Book TOC; 'Track-at-once' allows partial recording of the disk, with the option to record more at a later time, but without the option to read any of the data back until the disk TOC is fixed; 'Multisession' allows partial recording of the disk to a total of 99 sessions, with the option to read back the recorded data before the disk has been filled (provided that the reading drive is multisession capable, and can read the temporary Orange Book TOC).

The CD-R is intended for purposes which require CD compatibility rather than fast access times and updatability. As such it is not immediately suitable for real-time audio editing and production, but can be seen to have considerable value for the storage and transfer of sound material which does not require real-time editing. Broadcasters see an immediate usage in sound effects libraries, since a professional CD player may be cued up and played-in manually, or under external control, with sufficient accuracy for this purpose. Studios and mastering facilities may use them for providing customers and record companies with 'acetates' or test pressings of a new recording, and they have become quite popular as a means of transferring finished masters to a CD pressing plant in the form of the PMCD (pre-master CD). They are ideal as a means of 'proofing' CD-ROMs and other CD formats, and can be used as low cost backup storage for computer data.

4.5.8 The Sony Master Disc Format

Sony recently launched its own recording and editing system based on magneto-optical disks, called the 'Master Disc' format, which although physically resembling a conventional ISO standard 5.25 inch optical cartridge is a slightly different shape and not compatible (see Figure 4.14). The Master Disc is not a form of Compact Disc, and is not compatible with any type of CD-R machine on the market. The Master Disc format is marketed very much as a dedicated audio recording format rather than a computer hard disk, but it has many of the features of conventional computer hard disks.

Figure 4.14 Photograph of a Sony MasterDisc. (Courtesy of Sony Broadcast and Professional Europe)

The company has opted not to use conventional ISO computer M-O (magneto-optical) cartridges in this system. The Master Disc is single-sided and offers around 1.3 Gbytes of storage on that single side, suggesting a considerably higher data density on the disk than a typical M-O. The reason for adopting such a disk is clear when one realises that it can store up to 80 minutes of stereo audio at 20 bit resolution, using a 44.1 kHz sampling rate. When the audio is sampled at 16 bits the disk will store around 100 minutes of stereo, and at 24 bits will hold some 60 minutes. This makes it possible to master a complete CD on a single disk, and thus shows the format as directly aimed at the CD mastering market; but in contrast to older 16 bit mastering equipment the new format allows mastering at 20 bit resolution, or even 24 bits if future applications require it.

Unlike most M-O disks, the Sony disk requires no erase pass before rerecording over old material. In this sense it is similar to the consumer MiniDisc, and indeed this format seems to owe a considerable amount to the MiniDisc, being in many ways like a larger version of the same disk (except that MiniDisc stores compressed audio data and Master Disc stores linear PCM). Since rerecording is achieved by direct overwriting of old material, recording requires no more disk activity than replay, making possible various useful professional modes of operation such as read-after-write (confidence replay) and synchronous drop-in using read–write–read.

The blank disks are pre-grooved, with fixed address marks for sectoring. Within the format there is room for a certain amount of auxiliary data, including a timecode channel which may be asynchronous with the audio sample rate if required, and it is thus possible to include PQ data for CD mastering either in a burst at the start of the recording (as with U-matic masters) or alongside the audio in real-time (as on a consumer CD). Master Discs can be transferred at over twice normal play speed, using a SCSI bus.

4.6 Tape storage media

There are currently four main types of storage media in common use for tape backup storage with AV workstations. All are cassette or cartridge formats. They are the DDS format (based on DAT drives), Exabyte, QIC and DLT. DDS and Exabyte are by far the most popular with AV workstations at the moment because the media are widely available, low cost and reasonably fast, but the relatively recent DLT format is likely to gain popularity owing to its high capacity and transfer rates.

4.6.1 DDS

DDS is the DAT Data Storage format, and rather like the CD-ROM is the extension of a format originally intended purely for audio to general purpose data storage applications. The DAT format uses 4 mm tape and the tape is read and written using heads mounted in a drum which scans the tape in a helical fashion. On top of the audio DAT formatting is added formatting and error correction information so that the tape is then useful as a block-structured medium with low enough error rates for data purposes, and a directory area at the start of the tape.

DDS drives normally have four heads on the drum so that the data can be verified immediately after it is written – important for checking data reliability. It is recommended that one uses special DDS tapes for data purposes, which are said to be manufactured to the high specifications needed to ensure reliability, but some users have been known to use audio DAT tapes with varying degrees of success. It is sometimes necessary to alter a switch inside the drive for this purpose, so that it accepts ordinary tapes. It is important not to confuse audio DAT and DDS. This sometimes happens because it is possible to back up some AV workstations to an audio DAT machine using a digital audio interface to transfer files in real time, coupled with edit list information in the form of modulated data. Having said this, an example now exists of a drive which will act both as an audio DAT and a DDS drive, giving maximum flexibility in operation.

DDS-1 drives store up to 2 Gbytes of data on a tape, and some drives incorporate built-in data compression which can boost the storage capacity of such drives up to a maximum of 8 GBytes. This is lossless compression (see section 3.2) and so it allows the data to be recovered in precisely its original form. The degree of compression achievable depends on the type of data involved, the 8 Gbyte limit being the maximum achievable. Using compression slows down the storage process but increases the available capacity. The transfer rate to and from a DDS-1 drive is moderate (of the order of 180 kbyte/s), and the access time is quite slow compared with a disk drive (of the order of seconds).

DDS-2 drives offer higher storage capacity and higher transfer rates. Using a longer tape, the DDS-2 drive can store up to 4 Gbytes of data in uncompressed form, and up to 16 Gbytes compressed. The transfer rate is approximately 500 kbyte/s.

4.6.2 Exabyte

Exabyte tapes are based on the original consumer Video-8 format, adapted for data storage. The tapes are 8 mm wide, as opposed to the 4 mm of DDS, and the cartridges are slightly larger. Drives are typically more expensive than DDS drives. Storage capacities and transfer rates available from Exabyte drives are considerably greater than those available from DAT. One current example holds up to 5 Gbytes per tape and transfers data at a rate of around 500 kbyte/s. Maximum available capacity is currently 7 Gbytes. Exabyte tapes are becoming widely used as the medium on which CD masters are transferred from mastering houses to pressing plants, using DDP (Disk Description Protocol) files. The capacity of Exabyte and its ability to transfer CD masters a number of times faster than real-time makes it ideal for this purpose.

4.6.3 QIC

The QIC (quarter-inch cartridge) is quite a well established tape backup medium, used widely in professional computing and mainframe systems. It uses quarter-inch tape housed in a largish cartridge, and has very low error rates and high longevity. Recording is via stationary heads with multiple narrow tracks. Capacities and transfer rates are quite high, with drives storing over 10 Gbytes planned.

4.6.4 DLT

Digital Linear Tape (DLT) drives use a large number of linear tracks (128) across the width of a half-inch tape. The performance of DLT drives currently exceeds that of any of the other formats described above, with uncompressed capacity of up to 20 Gbytes available. Using a SCSI-2 interface, these drives offer transfer rates of up to 1.5 Mbyte/s with very low error rate, which makes them ideal for workstation backup purposes.

4.7 SCSI

By far the most commonly used interface for connecting mass storage media to host computers is SCSI (the Small Computer Systems Interface), pronounced 'scuzzy' by those in the know. This is a high speed parallel interface found on many computer systems, and allows up to seven peripheral devices to be connected on a single bus. Such peripheral devices include all forms of mass storage media, CD drives, scanners, printers and network ports. It is specified in ANSI X3.131 (1986). More recently a SCSI-2 standard has been developed which can be both faster and wider than SCSI-1, allowing for higher speed data transfer (SCSI-1 interfaces were limited to speeds of around 4–5 MByte/s, and were only 8 bits wide, whereas SCSI-2 can run at over 10 MByte/s and may be 16 or even 32 bits wide). The speed of SCSI interfaces is not fixed, but varies depending on the device and its software. Indeed the interface can be slower than many disk drives, making it into something of a bottleneck in some applications. This makes it advisable not only to check the speed of any disk drive to be used but also to check the SCSI port transfer rate of both disk drive and host computer or SCSI card. There is no point having a super fast disk drive if the SCSI interface cannot handle data at that rate.

The SCSI bus is the root of many of the problems encountered with storage media, and is known to be temperamental, especially when used with low cost cables and connectors. Some basic guidelines should help the most straightforward problems to be avoided.

4.7.1 SCSI addressing

SCSI devices are connected in a 'daisy-chain' fashion, as shown in Figure 4.15. SCSI-1 devices have two 50-pin connectors for this purpose, although some computers like the Macintosh have a non-standard 25-pin D-type connector. All devices on the SCSI bus see the same information: if one device sends something then all the others see it, leading to the need for device addresses to determine the transmitter and receiver of data. SCSI devices all have a means of setting their address, either with a DIP switch, a rotary or push button switch, and this determines the address on which the device will respond. The highest numbered address has the highest priority on the bus, and will be dealt with first, which helps when two devices conflict in attempting to access the bus.

Figure 4.15 Interconnection of SCSI devices

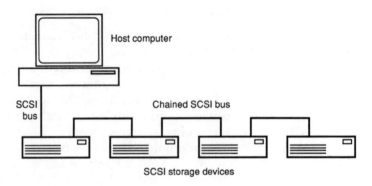

only to check the speed of any disk drive to be used but also to
Host computer

SCSI bus

Chained SCSI bus

SCSI storage devices

Normally the host computer has the highest address (ID7), leaving ID0 through ID6 for peripherals. A computer's internal hard disk often uses ID0. It is important to ensure that all devices on the bus have different addresses, otherwise problems arise, although it is not necessary to assign SCSI IDs in sequence.

Figure 4.16 Termination of a SCSI chain, showing use of an external terminator on the last device in the chain

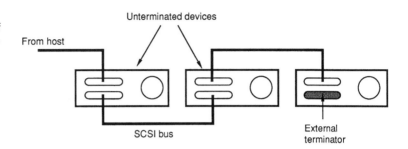

4.7.2 SCSI termination

The SCSI bus requires termination at both ends (one end is normally in the host computer or card and is not modifiable). This termination is a collection of resistors connected to each of the parallel lines which ensure that the termination impedance of the bus is correct, in order that the data is not distorted by reflections or attenuated. Unterminated SCSI buses occasionally work, but it is not recommended.

Termination can be either internal or external to the peripheral, and it may be switchable or even automatically sensed and controlled. Internal unswitchable termination is not advisable because it forces one to use the terminated device at the end of the SCSI chain (see Figure 4.16). It is particularly inconvenient if more than one SCSI device is to be connected, because the termination has to be physically removed from those devices in the middle of the chain (not always easy). External termination normally involves plugging a termination block into the daisy-chain connector of the last device in the chain. These can be easily purchased from computer stores. Automatic termination is useful because it means that the user does not need to think about which devices are in which positions on the bus – the device senses the impedance of the bus and terminates or not accordingly. Only the devices at each end of the bus should be terminated, not any of those in between.

4.7.3 SCSI cables

'The shorter the better' is the motto when it comes to choosing cables. Data rates are very high on the SCSI bus, and it is important to limit cable lengths to less than a metre where possible, otherwise errors will arise. The maximum total cable length allowable is 7 metres, but this depends on using high quality cables which are double-screened. Poor quality cables are the root of many problems encountered with SCSI buses.

4.7.4 Typical SCSI problems

The most common problems to arise involve (a) computers failing to 'see' certain peripherals; (b) systems failing to boot up properly; (c) data errors resulting in erroneous file transfers; (d) system crashes and 'glitches'. The following hints form a first-level troubleshooting guide:

- Never connect or disconnect devices with power turned on.
- Check that all devices have different addresses.
- Check all cables and connectors for soundness.
- Try swapping cables around or changing cables.
- Try shorter cables.
- Check termination and change if necessary.
- Try putting devices in different physical positions in the chain.
- Try changing the order of SCSI addresses.
- Try powering up SCSI devices in a different order.
- Try moving devices apart physically.
- Ensure that the correct device drivers are installed on the host computer.
- Run a SCSI diagnostic software tool which may point to the fault.

4.8 PCMCIA

PCMCIA is a standard expansion port for notebook computers and other small-size computer products. A number of storage media and other peripherals are available in PCMCIA format, and these include flash memory cards, modem interfaces and super-small hard disk drives. The standard is of greatest use in portable and mobile applications where limited space is available for peripheral storage.

4.9 Filing systems and volume partitions

So far only the physical structure and basic format of mass storage have been described. The way in which this raw storage space is used is another issue altogether. There are a number of ways of organising the storage capacity of a disk drive which involve formatting it at a high level for a particular filing system, depending on the computer platform or other host device and its operating system. It is this which determines whether the data stored on a disk or tape will be recognisable to the host computer, and whether it will be accessible to it.

When a disk is formatted at a low level the sector headers are written and the bad blocks mapped out. A map is kept of the

Figure 4.17 A disk may be divided up into a number of different partitions, each acting as an independent volume of information

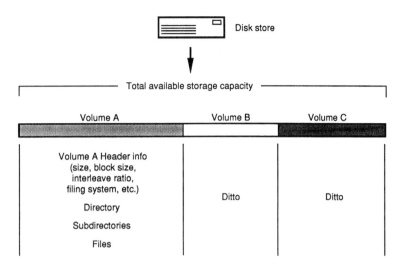

locations of bad blocks so that they may be avoided in subsequent storage operations. During a high level format the disk may be subdivided into a number of 'partitions'. Each of these partitions may behave as an entirely independent 'volume' of information, as if it were a separate disk drive (see Figure 4.17). It may even be possible to format each partition in a different way, such that a different filing system may be used for each partition. Each volume then has a directory created, which is an area of storage set aside to contain information about the contents of the disk. The directory indicates the locations of the files, their sizes, and various other vital statistics.

The file is the smallest unit of the filing structure. It is a simple container for data which usually begins with a header of some sort to indicate the type of file. A file is made up of a series of bytes of data which must be read in sequence to reassemble the data. Many filing systems store files in transfer blocks which are larger than the size of a single disk sector. Generally the larger the volume size the larger the size of the transfer block, since there is normally a logical limit to the number of addressable units in the filing system. A file is obliged to occupy an integer number of blocks, even if it does not need all of the last one, and this can lead to inefficient use of storage space if a lot of small files are stored. Large blocks are more useful for audio and video systems, because they improve transfer rates and the files are large anyway. A volume with a lot of small files can take up more space if copied directly to another volume which uses larger blocks, because the space is used less efficiently.

Figure 4.18 Disk files are organised in a tree-like structure, as shown in this example. Each directory can have subdirectories, down to many levels, within which may be files or further directories

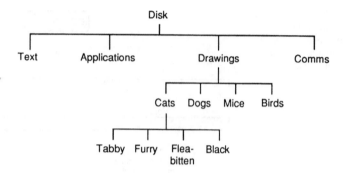

Most modern filing systems are hierarchical. In other words they are in a form of pyramid or tree structure which allows one master directory to contain a number of sub-directories, each of which may themselves contain sub-directories, as shown in Figure 4.18. In graphical user interfaces (GUIs), such as the Macintosh or Windows, these directories are often shown as 'folders', within which are either further folders or files. In text-based operating systems one must issue text commands to pick the directory by name. This is just a convenient way of organising the disk contents so that a user can group files by subject or any other useful method.

The most common filing systems in desktop computer-based AV workstations are the Macintosh HFS (Hierarchical Filing System) and the MS-DOS filing system (used on PCs). The Unix operating system is used on multi-user systems and some high-powered workstations, and it has its own filing system, but it is not commonly used for audio and video purposes. These though were not designed principally with real-time requirements such as audio and video replay in mind, and there are more suitable approaches for multimedia systems. There is, though, the advantage that disks formatted for a widely used filing system will be more easily interchangeable than those using proprietary systems. A number of workstation manufacturers have developed their own filing systems which are optimised for speed and efficiency in real-time applications. In many cases this has been the key to their success because it has allowed them to obtain more simultaneous audio channels from a given disk than would otherwise have been possible. Sonic Solutions, for example, has developed the Media Optimised File System (MOFS) for this purpose.[2]

It is possible, now that disk drives have become cheap and fast, that the need for special filing systems will become less important, and compatibility will become more important because of the need to interchange data between systems. There are, though, software means of allowing one system to interpret another's filing system, either directly or over a network. It is worth noting that some of the world's most widely used audio workstations use the ordinary Macintosh HFS for storing audio and compressed video.

4.10 Compatibility issues with mass storage media

There are a number of levels at which compatibility must exist before a disk or tape from one system can be read and interpreted by another. Most people know that MS-DOS computers won't read Macintosh disks, for example, and that Macs can only be made to read MS-DOS disks with the addition of some useful software such as PC Exchange. This is an example of the problem. Operating systems format disks in different ways, and this relates not particularly to the data files on the disks but to the way that the sectors are formatted, files stored, and the directory structured. This is the reason why you can't just take a SCSI disk which previously had been connected to one workstation and connect it to a different type and expect there to be no problems. Electrically there would be no problem (both could use a SCSI interface) but one's operating system might not be able to interpret the disk format of the other. Similarly with removable media, where there is the additional problem of physical compatibility. Figure 4.19 shows the hierarchy of levels at which compatibility may need to exist.

It is possible to equip workstations with drivers that will read and write disks in filing systems other than their own, but the need to do so has to be recognised and the solution may not be immediately straightforward. File names in one system may be restricted in a different way to those in others, for example. It is clear that although it might be convenient for audio file exchange to be tied to certain specific physical media, filing structures or network protocols, this is not really in the best long-term interests of the audio industry. The disks and tapes that are used are only ways of getting formatted data from place to place, and we may expect to change these physical interchange media quite regularly as the years go by, in order to take advantage of speed and capacity increases, as well as improvements in open systems design that may come along.

It is important to separate discussion of physical media from discussion of data formats and filing systems. One physical

Figure 4.19 Levels of compatibility in removable media interchange

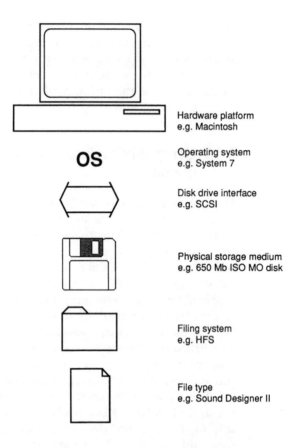

Hardware platform
e.g. Macintosh

Operating system
e.g. System 7

Disk drive interface
e.g. SCSI

Physical storage medium
e.g. 650 Mb ISO MO disk

Filing system
e.g. HFS

File type
e.g. Sound Designer II

medium could be formatted in a whole variety of different ways. This is a departure from the old ways of the audio and video industries in which dedicated formats existed which specified dedicated physical media. An audio DAT tape, for example, should be playable in any audio DAT machine, but a computer M-O disk with some sound files on it is much more difficult to deal with because it depends on all the factors indicated above.

The issue of open systems interchange is discussed in greater detail in Sections 6.1 and 6.2.

4.11 Formatting, fragmentation and optimisation of media

The process of formatting a disk or tape erases all of the information in the volume. (It may not actually do this, but it rewrites the directory and volume map information to make it seem as if the disk is empty again.) Effectively the volume then becomes virgin territory again, and data can be written anywhere.

Figure 4.20 At (a) a file is stored in three contiguous blocks and these may be read sequentially without moving the head. At (b) the file is fragmented and is distributed over three remote blocks, involving movement of the head to read it. The latter read operation will take more time

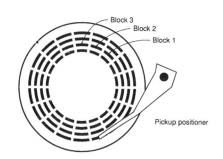

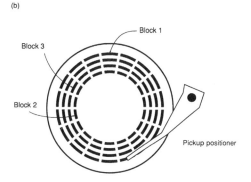

When an erasable volume like a hard disk has been used for some time there will be a lot of files on the disk, and probably a lot of small spaces where old files have been erased. New files must be stored in the available space, and this may involve splitting them up over the remaining smaller areas. This is known as disk fragmentation, and it seriously affects the overall performance of the drive. The reason is clear to see from Figure 4.20. More head seeks are required to access the blocks of a file than if they had been stored contiguously, and this slows down the average transfer rate considerably. It may come to a point where the drive is unable to supply data fast enough for the purpose.

There are only two solutions to this problem: one is to reformat the disk completely (which may be difficult, if one is in the middle of a project), the other is to optimise or consolidate the storage space. Various software utilities exist for this purpose, whose job is to consolidate all the little areas of free space into fewer larger areas. They do this by juggling the blocks of files between disk areas and temporary RAM – a process which often takes a number of hours. Power failure during such an optimisation process can result in total corruption of the drive, because the job is not completed and files may be only half moved, so it is advisable to back up the drive before doing this. It has been known for some such utilities to make the files unusable by some audio editing packages, because the software may have relied on certain files being in certain physical places, so it is wise to check first with the manufacturer.

4.12 Magnetic tape interchange

Tapes are not usually formatted in the same way as disks. Tapes are often used as basic 'data streamers' where data is stored in a very simple sequential fashion, possibly even with the block size varying in different parts of the tape. It may be that no directory is stored on the tape itself, this being kept in a disk file on the host computer. An ANSI standard exists[3] which defines basic rules for information interchange on magnetic tapes, and this is often used on media such as Exabyte to determine the method of labelling tapes and filing information. Because tapes are not usually 'mountable volumes' in the same way as disks, it is rare to be able to 'see' them on the desktops of GUI-based computers, requiring special software with appropriate drivers for the tape system in question to read and write information.

References

1   Philips (1995) *Enhanced Music CD Specification Version 0.9*. Philips Consumer Electronics, Building SWA-1, PO Box 80002, 5600 JB Eindhoven, The Netherlands.
2   Anderson, D. (1993) 'High speed networking for professional digital audio'. Presented at AES UK Digital Audio Interchange Conference, 18–19 May, London, p. 67. Audio Engineering Society.
3   ANSI (1978) *X3.27-1978: Magnetic tape labels and file structure for information interchange*. American National Standards Institute.

5 Audio handling in the workstation

This chapter explains the ways in which digital audio is stored, replayed and manipulated in digital audio workstations, using examples from real systems.

5.1 Anatomy of a digital audio workstation

There are fundamentally two ways of putting together an audio workstation: either build a dedicated system or install additional hardware and software into a standard desktop computer. Most systems conform to one or other of these models, with a tendency for dedicated systems to be more expensive than desktop computer-based systems (although this is by no means always the case).

5.1.1 Dedicated systems

It was most common in the early days of hard disk audio systems for manufacturers to develop dedicated systems with fairly high prices. This was mainly because mass produced desktop computers were insufficiently equipped for the purpose, and because large capacity mass storage media were less widely available than they are now, having a variety of different interfaces and requiring proprietary file storage strategies (see section 5.2). It was also because the size of the market was relatively small to begin with, and considerable R&D investment had to be recouped.

Figure 5.1 A dedicated digital audio workstation. (Courtesy of Digital Audio Research)

There are considerable advantages to dedicated systems, and they are very popular with professional facilities. Rather than a mouse and a QWERTY keyboard, the user controls the system using an interface designed specifically for the purpose, with perhaps more ergonomically appropriate devices. An example of such a system is pictured in Figure 5.1. It has a touch screen and dedicated controls for many functions, as well as rotary and slider controls for continuously variable functions. It is also becoming common for cheaper dedicated editing systems to be provided with an interface to a host computer so that more comprehensive display and control facilities can be provided.

5.1.2 Desktop computer-based systems

In recent years desktop multimedia computers have been introduced with built-in basic AV (audio-video) facilities, providing limited capabilities for editing and sound manipulation. The quality of the built-in convertors in desktop computers is necessarily limited by price, but they are capable of 16 bit, 44.1 kHz audio operation in many cases. Better audio quality is achieved by using third party hardware.

Many desktop computers lack the processing power to handle digital audio and video directly, but by adding third-party hardware and software it is possible to turn a desktop computer into an AV workstation, capable of storing audio for an almost unlimited number of tracks with digital video alongside. An audio signal processing card is normally installed in an expansion slot of the computer, as shown in Figure 2.8. The card would be used to handle all sound editing and post-processing operations, using one or more DSP chips (see section 2.8), with the host computer acting mainly as a user interface. The audio card would normally be connected to an audio interface,

Figure 5.2 Typical system layout of a desktop computer based audio workstation

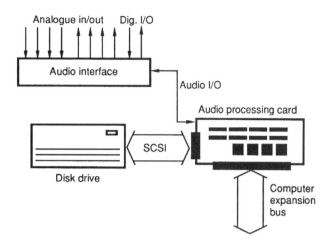

perhaps containing a number of A/D and D/A convertors, digital audio interfaces such as AES/EBU (see Chapter 6), probably a SMPTE/EBU timecode interface (see section 7.6), and in some cases a MIDI interface (see section 8.3). A SCSI interface to one or more disk drives is often provided on the audio expansion card in order to optimise audio file transfer operations, although some basic systems use the computer's own SCSI bus for this purpose. The photographs in Figure 5.3 show the expansion cards for a popular system of this type.

To date, the Apple Macintosh computer has been the most popular computer for use as an audio workstation. It uses a high

Figure 5.3 Audio processing cards for a desktop PC. (Courtesy of Studio Audio Video)

speed internal expansion bus (the NuBus), has a good graphical user interface and has used the SCSI bus for interface to peripherals since 1986. In recent years a number of products based around the MS-DOS PC and the Windows operating systems have been introduced, as well as miscellaneous systems on other platforms such as the Unix-based SGI. Both platforms are now intending to use the so-called PCI expansion bus, replacing Apple's NuBus, and system software from one platform can increasingly be emulated on the other if desired, making the distinction between hardware platforms less clear.

5.2 Principles of sound recording on mass storage media

5.2.1 The sound file

In digital audio workstations, recordings are stored in sound files on mass storage media. The storage medium is normally a disk, but other media may be used in certain circumstances such as for backup of disks. For the sake of simplicity disks are assumed to be the primary means of storage in the following sections. A sound file is an individual recording of any length from seconds to hours (within the limits of the system). With tape recording, parts of a tape may be recorded at different times, and in such a situation there will be sections of that tape which represent distinctly separate recordings: they may be 'tracks' for an album, 'takes' of a recording session, or short individual sounds such as sound effects. This is the closest that tape recording gets to the concept of the 'sound file': that is a distinct unit of recorded audio, the size of the unit being anything which fits into the available space.

In the audio workstation one must consider the disk as a 'sound store' in which no one part has any specific time relationship to any other part – no section can be said to be 'before' another or 'after' another. This is the nature of random- or direct-access storage (although note that some forms of optical disk, such as the WORM cartridge and CD-R record contiguously for all or part of their capacity, whilst retaining random accessibility). It has led to the use of the somewhat confusing description 'non-linear recording', which contrasts with the 'linear' recording process that takes place on tape. (To many people the term 'non-linear' means that the audio has been quantised non-linearly, which is not the case in most professional audio workstations.)

A disk may accommodate a number of sound files of different lengths. It is possible that one file might be a 10 minute music track whilst another might be a 1 second sound effect. Essentially, one may keep as many sound files in the store as will fit in the space available, although some operating systems have upper limits on the number of individual files that can be handled by the directory structure. Each sound file is made up of a number of discrete data blocks, and normally the block size will limit the minimum size occupied by a file since systems do not normally write partial blocks (see below).

Normally sound files are either mono or stereo – that is either a single channel or two related channels of audio combined into one file. They are rarely more than stereo, since multichannel operation is normally achieved by storing a number of separate mono files, one for each channel. Stereo sound files contain the left and right channels of a stereo pair, usually interleaved on a sample-by-sample basis as described in section 6.1, and are useful when the two channels will always be replayed together and in a fixed timing relationship. Accessing a stereo file is then no different from accessing a mono file, except that the stereo file requires twice the amount of data to be transferred for the same duration of audio. As far as the user is concerned, the system can present a stereo sound file under a single title and note in the file header that it is stereo. In this case any buffering (see below) would have to be split such that left channel samples would be written to and read from one group of memory addresses, and right channel samples to and from another. As would be expected, stereo files take up twice the amount of disk space of the equivalent mono file.

5.2.2 RAM buffering

Computer disk drives were not originally designed for recording audio, although they can be made to serve this purpose. As shown in the previous chapter, a disk is normally formatted in sectors, often grouped into blocks, and the blocks making up a file may not be stored contiguously (contiguous means physically adjacent). The result of this is that data transfer to and from such media is not smooth but intermittent or burst-like. Furthermore, editing may involve the joining of sections from files stored in physically separate locations, resulting in breaks in the data flow from disk at edit points whilst the new file is located. Although this burst transfer rarely presents a problem in applications such as text processing (it does not matter if a text file is loaded in bursts) it is unsuitable for the recording and replay of real-time audio. Audio (and video in most cases) requires that samples are transferred to and from convertors or

Figure 5.4 RAM buffering is used to convert burst data flow to continuous data flow, and vice versa

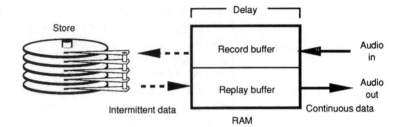

digital interfaces at a constant rate, in an unbroken stream. Consequently digital audio hardware and software must include mechanisms for converting burst data flow into continuous data flow, and vice versa, and this is achieved by using RAM as a short-term 'buffer' or reservoir.

RAM (see section 1.6) is temporary solid-state memory with a very fast access time and transfer rate. It can be addressed directly by the CPU and DSP hardware of the audio workstation, and is used as an intermediary store for audio samples on their way to and from the disk drive (see Figure 5.4). During recording, audio samples are written into the RAM at a regular rate and read out again a short time later to be written as blocks of data on the disk. At least one complete sector of audio is transferred in one operation, and usually a number of sectors are written in one operation (see section 5.2.4). The transfer is effectively time compressed, since samples acquired over, say, 100 ms, may be written to the disk in a short burst lasting only 20 ms, followed by a gap. During simple replay, data blocks are transferred from the disk into RAM in bursts and then read out at a steady rate for transfer to a D/A convertor or digital interface. The process of transferring out from the buffer normally begins before the file has been transferred completely into the buffer, because otherwise (a) there would be an unacceptable delay between the initiation of replay and the onset of an audible output, and (b) the size of the buffer would have to be great enough to hold the largest sound file entirely.

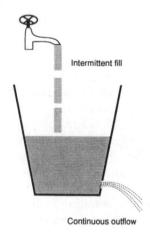

Figure 5.5 RAM buffering may be likened to a water reservoir which acts to convert intermittent filling to continuous outflow

The RAM buffer acts in a similar way to a water reservoir. It allows supply and demand to vary at its input and its output whilst remaining able to provide an unbroken supply, assuming that sufficient water remains in the reservoir. Figure 5.5 shows an analogy with a water bucket that has a hole in the bottom, filled by a tap. One may liken the tap to a disk drive and the water flowing out of the hole to an audio output. The tap may fill the bucket in bursts, but within certain limits this is converted into continuous outflow. Provided that the average

Figure 5.6 A control system could be added to the simple reservoir to regulate inflow and outflow so that supply and demand were linked

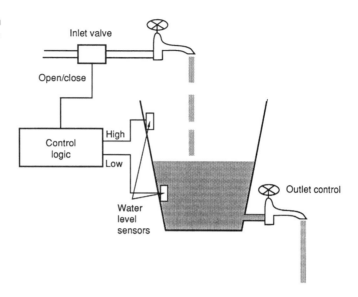

flow rate of water entering the bucket is the same as the average rate at which it flows out of the hole, then the bucket will neither empty nor overflow (within the limits of the size of the bucket). If water flows out of the hole faster than it is supplied by the tap then the bucket will eventually become empty. On the other hand, the bucket could overflow if the tap was left on all the time and was filling the bucket faster than the hole could empty it.

Clearly some control mechanism is called for. Sensors could be attached to the insides of the bucket to detect high and low water levels, as shown in Figure 5.6, connected to control logic which operated a valve in the supply line. The valve would be opened when the water level was getting low, and closed when it was getting high. A tap on the bucket outlet could be added to stop and start the flow (the equivalent of the PLAY button for audio replay). Equivalents of this control mechanism exist in audio workstation software. Pointers are incremented up and down to register the state of fullness of RAM buffers during record and replay operations, and action is taken at certain states of fullness either to transfer new blocks of data to and from the disk or to halt transfer.

The analogy can be taken further. There might be more than one hole in the bucket (more audio outputs), larger holes in the bucket (higher sampling rates and resolutions) or a tap with low water pressure (a slow storage device). Audio system design is largely a matter of juggling with these parameters and others to optimise the system performance. (The bucket analogy does not hold water if examined too closely, as water will flow faster out of the holes in the bucket the fuller the bucket, and this does not hold true for memory buffers in audio workstations!)

RAM buffering has a number of other uses. First, it can be used to ensure that any short-term timing irregularities in the data coming from the storage device will be ironed out and will not be allowed to affect audio quality. Data written into memory from the store, even if it has timing jitter, can be read out from the store at a constant steady rate, under control of an accurate crystal clock. The only penalty of buffering is that it introduces a small delay between the input to and the output from the buffer, the extent of which depends on the delay between the writing of samples to the RAM and the reading of them out again. The maximum delay is limited by the size of the buffer, as with a small buffer there will come a point where the memory is filled and must be partially emptied before any new samples can be written in. The delay effect of the buffer can be disguised in operation because data can be read from disk ahead of the required time, and written at an appropriate time after sample acquisition.

Secondly, the buffer may be used for synchronisation purposes. If audio data is to be synchronised with an external reference such as a timecode, then the rate at which data is read out of the buffer can be finely adjusted to ensure that lock is maintained. In practice the process of ensuring synchronisation without the audible side effects of timing jitter is a more complex subject, and this will be discussed further in section 5.5.

The size of buffer in a digital audio system may or may not be under the user's control, but is typically in the region of 0.5–2 Mbytes. An area of operating RAM will be set aside for this purpose, often located on the audio processing board itself rather than being system RAM of the host computer. Generally, the more channels to be handled, the larger the buffer, since each channel requires its own memory space; also a larger buffer can help to compensate for badly fragmented storage space (see section 4.11), although it cannot make up for a disk drive which is too slow overall.

5.2.3 Disk drive performance issues

In Chapter 4 the most important performance characteristics of mass storage media were introduced. Access time and transfer rate were found to be important features governing the suitability of disk drives for primary digital audio storage. It was seen that the sustained transfer rate was a far more important feature than the instantaneous rate, since this was more likely to represent the performance in real file transfer operations.

Table 4.1 and 4.2 showed the data rates and capacities required for different resolutions of digital audio, and from this one can begin to work out the performance requirements of storage devices. The data rate for one channel of audio at 48 kHz, 16 bits, amounts to around 0.75 Mbit/s, thus it might be assumed that a store with a transfer rate of 0.75 Mbit/s would be able to handle the replay of one audio channel's data satisfactorily. If the store was made up of solid state RAM which has a negligible access time (of the order of tens or hundreds of nano seconds) then a transfer rate of 0.75 Mbit/s would be adequate, but in the usual case where the store is a disk drive, the access time will severely limit the average transfer rate to and from the buffer. Although the burst transfer rate from the disk to the buffer may be high, the gaps between transfers as the drive searches for new blocks of data will reduce the effective rate. It is therefore the combination of access time and transfer rate that go to make up the effective transfer rate. What is needed is a fast transfer rate and a fast access time.

The job of the buffer is to disguise the effects of access time delays, and it may be seen that the size of the buffer will depend on the potential access delay, among other things. If transfer is erratic, that is with long gaps and then extremely fast transfers, the buffer is likely to swing between being very full and very empty, rather than deviating a small amount around a half full position. In the former case it is likely that a larger buffer will be required.

Over a period of time the disk is likely to become fragmented (see section 4.11) and this will lead to file blocks being stored in a number of divided locations. The more fragmented a store becomes the lower the efficiency of data retrieval, as a file will be transferred in a number of short bursts separated by breaks while the next block is accessed. Furthermore, the access time depends on how physically far apart the blocks are, as the retrieval mechanism will take less time to travel a physically short distance than to travel a long way. For this reason, figures quoted for access time can only ever be a rough guide.

Certain storage media have different access times and transfer rates when recording (writing) to those encountered when replaying (reading). For example, hard disks use a magnetic recording method which overwrites old information completely without erasing it first. Magneto-optical drives usually require a two stage process in order to rewrite over old data, in that the required block must be erased on one revolution and then written on the next, although there are various ways of tackling this limitation which were described in Chapter 4. There may also be a 'verify' pass after writing. This suggests that recording performance may not always be as good as replay performance, and that a disk drive may be able to replay more channels simultaneously than it can record.

For all the above reasons it is often difficult to calculate how many channels of audio one may expect a disk drive to be able to handle. To take an example, assume a disk drive with an average access time of 20 milliseconds and a transfer rate of 20 Mbit/s. If the access time was near zero then the transfer rate of 20 Mbit/s would allow around 26 channels of audio to be transferred at the example resolution given above, but the effective transfer rate in real operation will bring this number down to perhaps twelve or fewer channels for safe, reliable operation in a wide variety of operational circumstances. Editing, as described in section 5.4, also places considerable additional demands on disk drive performance, depending on how edits are carried out. Because of all this, some manufacturers play very safe and limit their systems to perhaps four or eight channels per disk drive, even if the drive might be able to handle more under some circumstances. They would rather have reliable performance all the time than 'sail too close to the wind'. The effect of using a disk drive beyond the limits of its performance is normally to experience 'drop-outs' in replay, and system messages such as 'drive too slow', when attempting to replay large numbers of channels with many edits.

5.2.4 Allocation units or transfer blocks

Optimising the efficiency of data transfer to and from a storage device will depend on keeping the number of head seeks to a minimum for any given file transfer, and this requires careful optimisation of the size and position of the audio transfer blocks or allocation units. Typically, a disk sector (that is the smallest addressable storage unit) contains 512 bytes of information, although some optical drives use 1024 byte sectors. This is very small in relation to the size of a digital audio file of even moderate length, and if a file were to be split up into chunks of 512 bytes spread all over the disk then efficiency would be impossibly reduced due to the large number of seeks required to different parts of the disk. For this reason a minimum transfer block is usually defined, which is a certain number of bytes that are transferred together and preferably stored contiguously in order to improve efficiency. It might be that a transfer block would contain 8 kbytes of audio data, which in the case of 512 byte sectors would correspond to sixteen sectors. The size of the transfer block must be small enough to engender efficient use of the disk space in cases of fragmentation, and large enough to result in efficient data transfer. If the digital audio system stores audio under the native filing system of the host computer then the size of the transfer block may be fixed during the formatting of the disk volume, as explained in Chapter 4.

5.3 Multichannel recording and replay

5.3.1 Multitrack or multichannel?

It is important to understand a fundamental difference between the workstation concept of multichannel operation and the conventional concept of multitrack recording as encountered with tape recorders. The difference is that 'tracks' and 'channels' need not necessarily mean the same thing. In a multitrack tape recorder, there may be perhaps up to 48 tracks of audio recorded onto the tape, each of which is an independent mono track lasting the length of the tape. Each of these independent tracks feeds a numbered audio output, and is fed from a numbered input. Once sound is recorded onto a numbered track it is fixed in time and physical position in relation to other sounds recorded on the same tape, and it will be replayed on the same-numbered audio channel at all times (unless the internal wiring of the machine is changed).

In workstations the terms 'track' and 'channel' may be separated from each other, in that a sound file, once stored, may be replayed on any audio channel depending on the user's choice. It may even be that the concept of the track is done away with altogether, but this depends on the user interface of the system, and most manufacturers have chosen to retain the concept of tracks because it is convenient and well understood. Tracks, in workstation terminology, are just ways of showing which sound elements have been grouped together for replay on the same channel, but they are not fixed as in tape recording. Figure 5.7 shows a simulated display from a multitrack package in which tracks are represented as horizontal bands containing sound file segments. On the left hand side it is possible to change the physical audio output assigned for replay of that track. The sound segments can be moved around in time on the virtual track by sliding them left or right, and they can be copied or moved to other tracks if necessary.

Figure 5.7 Tracks are represented in this simulated display as horizontal bands containing named sound file segments. The output to which that track is routed is selected at the left hand side, along with recording and replay muting controls

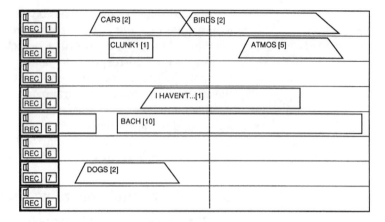

5.3.2 Inputs, outputs, tracks and channels

Because of the looser relationship between tracks, channels and audio inputs and outputs, confusion occasionally arises. First, it should be remembered that none of these are necessarily related to each other, although a designer may decide to relate them. In a twenty-four track tape machine, there are twenty-four inputs, twenty-four outputs, twenty-four tracks, and twenty-four channels, so it is very easy to see a direct relationship between one and the others. It is even possible to say exactly where on the tape track thirteen will be recorded at any point in time. In a workstation it is possible, for example, for there to be two inputs, eight outputs, 99 tracks, and eight channels. It is rarely possible to say exactly where track thirteen will be recorded at any point, or what information is recorded on it, as it all depends on what the user has decided. In this example it may be that only two inputs have been provided because that is all that the designer is going to allow you to record at any one time, but it is highly likely that these two inputs could be routed to any 'track' or any output channel. The two inputs allow for the recording of stereo or mono sound files which will be stored in a free location and given names by the user. Although only two 'tracks' may be recorded at once, this operation may be performed many times to build up a large number of sound files in the store.

In some systems, the concept of the track has been considered as important, and in the above example there are 99 tracks (just a virtual concept) but only eight outputs or channels. This is because the user is allowed to record information onto any of the tracks, but he may only replay eight of them simultaneously. The number of simultaneous output channels is limited by the transfer rates of the storage devices, the signal processing capacity of the system and the number of D/A convertors employed. By expanding the system, adding more or faster disks and adding more processing power, more of the 99 tracks could be replayed simultaneously. Many manufacturers have taken this modular approach to system design, allowing the user to start off in a small way, expanding the capabilities of the system as time and money allow.

5.3.3 Track usage, storage capacity and disk assignment

The storage space required for multiple channels increases pro rata with the number of channels, although in fact eight track recording may not require eight times the storage space of mono recording because many 'tracks' may be blank for large amounts of the time. If you think about the average multitrack recording on tape you will realise that many tracks have large gaps with nothing recorded. The total storage space used will depend on the total duration of the mono sound files used in the programme, whatever tracks or channels they are assigned to. Anderson[1] has estimated, for example, that sound effects tracks in feature film production contain about two-thirds silence, and that dialogue tracks are only 10–20 per cent utilised.

(It is often said that disk-based systems do not record the silences on tracks and therefore do not use up as much storage space as might be expected, but the only time when silences save storage time is when they exist as blank spaces between the output of sound files, where no sound file is assigned to play (see Figure 5.8). Recorded silence uses as much disk space as recorded music!)

Multichannel disk recording systems sometimes use more than one disk drive, and there is a limit to the number of channels

Figure 5.8 The silent section on the upper track does not require any disk space because no recording exists for this time slot. The silent section on the lower track was recorded as part of a file, and so consumes as much space as any other sound

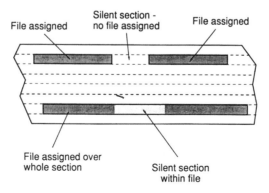

File assigned

Silent section - no file assigned

File assigned

File assigned over whole section

Silent section within file

Figure 5.9 Some older multitrack disk systems assigned disks permanently to certain tracks, as shown here

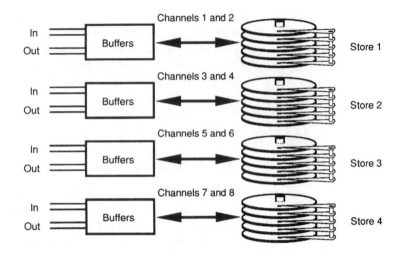

which can be serviced by a single drive. It is necessary, therefore, to determine first how many channels a storage device will handle realistically, and then to work out how many are needed to give the total capacity required. Some older systems attempted to imitate a multitrack tape recorder in assigning certain disk drives permanently to certain groups of tracks, as shown in Figure 5.9, but this limited operational flexibility. If a sound file from one track were needed on another it might have to be copied to the appropriate drive, which would take time. This approach is becoming much less common now that the performance of disk drives is getting to the point where one can replay perhaps 16 channels simultaneously from a single drive. In modular systems, if one needs greater channel recording and replay capacity one can simply add further disk I/O cards, each connected to a separate disk. If one needs more storage capacity then more disk drives can be attached to the same SCSI bus, as shown in Figure 5.10. It is then relatively unimportant which drive a file is stored on, provided that the software is capable of handling the addressing of multiple drives, although there may be some restrictions if the user has constructed a play list which requires more simultaneous file transfers from a certain disk than can be handled.

5.3.4 Multichannel storage and buffering

One of the principle features of non-linear recording is its ability to replay files to any output channel at any time. Preconceptions about fixed time relationships between files no longer hold true, and the order in which files are recorded need have no relation-

Figure 5.10 Arrangement of multiple disks in a typical modular system, showing how a number of disks can be attached to a single SCSI chain to increase storage capacity, and how additional disk I/O cards can be added to increase data throughput for additional audio channels

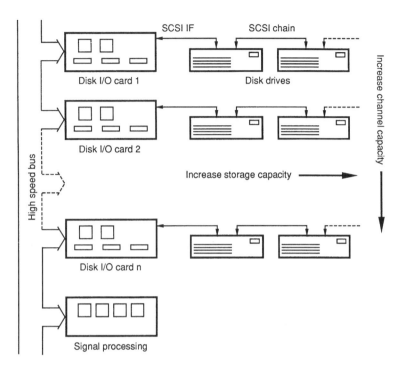

ship to the order in which they are replayed. To this can be added the fact that one disk drive may be supplying files for multiple outputs, so may in effect be required to access a number of files simultaneously. (Of course 'simultaneously' in this sense has to be a virtual concept, since the drive would in fact share its time between the different file transfers.) The idea that any file might be required at any time by the system, and routed to any output, is one reason why the physical location of the blocks making up files, and their relative 'nearness' to each other on the disk surfaces, may be less of a problem than at first appears. In multichannel operation, particularly for video and film post-production operations where small and large files are replayed in a variety of timing relationships, the head positioner on the audio disk drive is likely to be hopping about all over the place to service the requirements of each replay 'track'.

If the blocks of recordings which had originally been close in time were to be placed physically close on the storage device this would favour a particular timing relationship between them, as it would give an access time advantage in the case where those files were required together, but a disadvantage if one of them was needed at the same time as one which was physically far away. Because of this it might be suggested that any storage

147

strategy which places any one file physically closer to another than it is to the rest of the files in the store will favour a particular time relationship between those files. It would appear that the only viable strategy is in fact for the blocks to be randomly distributed throughout the store, in which case no one relationship is favoured over any other.

The technique of pseudo-random storage described above may optimise transfer efficiency in the case of true random access for a number of channels served by one drive, but in the case of a system attempting to imitate a multitrack tape machine, where only a few channels are served by each disk drive, it may be more sensible to adopt a strategy in which audio data is written to disk in a physically sequential order, such that a serial time relationship is established. This assumes that it will be the norm for files to be played throughout their lengths in a manner akin to tape replay. If such a situation is likely, then possibly the random storage strategy would result in lower efficiency.

In a multichannel system, where one storage device is being accessed to provide outputs for a number of channels, the system will be in the position of having to decide which output to serve at what time, and must keep the buffer for each channel at a reasonable state of fullness. This will be achieved in conjunction with a scheduled playing order of sound files which the user has decided upon, and which dictates which files are to go to which outputs at which times. The scheduling software will look at the user's playing order and determine what must be done in terms of file access in order to fulfil the order. Once the play-out process has begun the system will constantly refer to the state of emptiness of the buffers for each channel to see which needs filling the most, and will have to make decisions between either filling the emptiest buffer (otherwise it will cease to produce audio) or accessing blocks which are physically close in the store (giving the highest transfer efficiency).

Abbott[2] suggests that above a certain state of fullness of each channel's buffer the system should concentrate on accessing physically close blocks of data, as this is the most efficient in terms of transfer rate (see above), but that in the urgent situation where one or more channel's buffer is below a certain 'danger level' the system should concentrate on filling the emptiest buffer, ignoring the need for the most efficient transfer, until such time as the 'safe' situation is reached again when all channels are above the danger level. In this way a multichannel system will continually alternate between the 'safe' and the 'urgent' modes of operation depending on the state of emptiness

Figure 5.11 In (a) channel replay buffers are all moderately full, and thus the system may concentrate on accessing storage blocks which are physically close (for greater efficiency). In (b) two channel buffers are below the SAFE/URGENT boundary, and thus the system must concentrate on filling these at the expense of efficiency, otherwise they will become empty

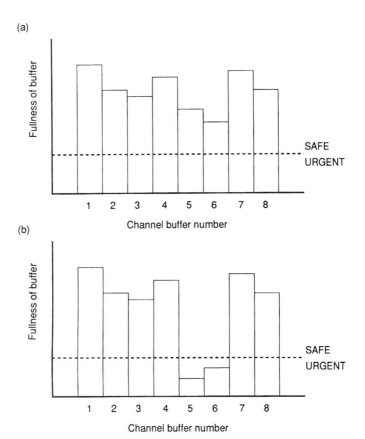

of each channel's buffer, and therefore maintain the most efficient transfer for any given playing schedule (see Figure 5.11).

5.3.5 Dropping-in

In multitrack music systems the capability to 'drop-in' is important. Dropping-in involves instantaneous entry into record mode at the touch of a button, and it is expected that a seamless join will result between old and new material, both at the start of the drop-in, and at the drop-out into the old material again.

Dropping in and out are really very similar operations to those involved in editing (see section 5.4), as a crossfade must be added between old and new material at the join. In terms of file operations, it may be appreciated that one cannot simply start to write new material half way through a previously written file, and thus it is necessary to write a new file for the 'dropped-in' portion. Internally, as part of the replay schedule, the system will then have to keep a record of times at which it must cross-fade from one file to the other and back again.

5.4 Audio editing

5.4.1 Advantages of non-linear editing

Speed and flexibility of editing is probably one of the greatest benefits obtained from non-linear recording. Tape editing has some advantages, but with digital audio it was often cumbersome, requiring material to be copied in real time from source tapes to a master tape, and presenting difficulties in making minor adjustments to a finished master. Tape-cut editing was very fast and cheap, being the main method used for years with analog tape, but it was rather unreliable on digital formats and little used in practice. Cassette tape formats like DAT were not designed with editing in mind, and cannot be cut. When cut-editing tape, the editor fixes the edited sections in a physical and therefore in a temporal relationship with each other. If he or she desires to change any aspect of the edited master then it must be taken apart and rejoined, and there will usually only be one final version of the master tape. With non-linear editing the editor may preview a number of possible masters in their entirety before deciding which should be the final one. Even after this, it is a simple matter to modify the edit list to update the master. Edits may also be previewed and experimented with in order to determine the most appropriate location and processing – an operation which is less easy with other forms of editing.

The majority of music editing is done today using digital audio workstations, indeed these are now taking over from dedicated audio editing systems because of the speed with which takes may be compared, crossfades modified, and adjustments made to equalisation and levels. Non-linear editing has also come to feature very widely in post-production for video and film, because it has a lot in common with film post-production techniques involving a number of independent mono sound reels.

Non-linear editing is truly non-destructive in that the edited master only exists as a series of instructions to replay certain parts of certain sound files at certain times, with certain signal processing overlaid, as shown in Figure 5.12. The original sound files remain intact at all times, and a single sound file can be used as many times as desired in different locations and on different tracks without the need for copying the actual audio

Figure 5.12 Instructions from an edit decision list (EDL) are used to control the replay of sound file segments from disk, which may be subjected to further processing (also under EDL control) before arriving at the audio outputs

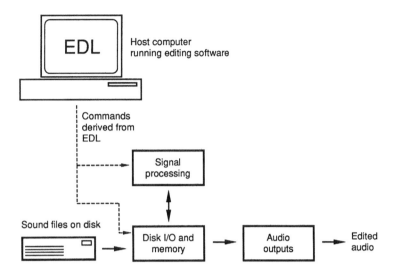

data. Editing may involve the simple joining of sections, or it may involve more complex operations such as long crossfades between one album track and the next, or gain offsets between one section and another. The beauty of non-linear editing is that all these things are possible without in any way affecting the original source material.

The process of editing involves the regular comparison of different 'takes' to determine the suitability of each, and with tape this can be a time-consuming process due to the need for spooling of tapes to possibly distant locations, or even the changing of reels. In the non-linear system the comparison of takes need take no more than a few seconds, and it may be possible to address the system in terms of the real title of the desired section, whereupon this section will be found, instead of needing to refer to counter or timecode locations logged during a session.

5.4.2 Sound files and sound segments

Sound files have already been defined in section 5.2.1: they are the individual sound recordings contained on a disk, each of which is catalogued in the disk directory. In the case of music editing sound files might be session takes, anything from a few bars to a whole movement, and in picture dubbing could be a phrase of dialogue or a sound effect. They are normally stored with a name to identify them. During editing one may isolate certain segments of these sound files which are useful in themselves, because it may be that one wishes to use only a portion of a sound file rather than the whole file. In such cases

151

it is useful to be able to identify the segment as an entity in its own right, so that it can be named and used wherever required, and this is a feature provided in most editing systems.

Rather than creating a copy of the segment and storing it as a separate sound file, it is normal simply to store the segment as a 'soft' entity – in other words as simply commands in an edit list or session file which identify the start and end addresses of the segment concerned, and the sound file from which it came. It may perhaps be given a name by the operator and subsequently used as if it were a sound file in its own right. An almost unlimited number of these segments can be created from original sound files, without the need for any additional audio storage space.

5.4.3 Edit point handling

Edit points can be simple butt joins or crossfades. A butt join is very simple because it involves straightforward switching from the replay of one sound segment to another. Since replay involves temporary storage of the sound file blocks in RAM (see above) it is a relatively simple matter to ensure that both outgoing and incoming files in the region of the edit are available in RAM simultaneously (in different address areas). Up until the edit, blocks of the outgoing file are read from the disk into RAM and thence to the audio outputs. As the edit point is reached a switch occurs between outgoing and incoming material by instituting a jump in the memory read address corresponding to the start of the incoming material. Replay then continues by reading subsequent blocks from the incoming sound file. It is normally

Figure 5.13 (a) A bad butt edit results in a waveform discontinuity. (b) Butt edits can be made to work if there is minimal discontinuity

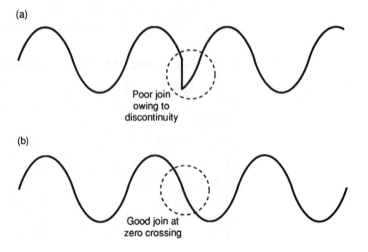

(a)

Poor join owing to discontinuity

(b)

Good join at zero crossing

possible to position edits right down to single sample accuracy, making the timing resolution as fine as a number of tens of microseconds if required.

The problem with butt joins is that they are quite unsubtle. Audible clicks and bumps may result because of the discontinuity in the waveform that may result, as shown in Figure 5.13. It is normal, therefore, to use at least a short crossfade at edit points to hide the effect of the join. This is in fact what happens when analog tape is spliced, since an angled cut is used which has the same effect as a short crossfade (of between 5 and 20 ms depending on the tape speed and angle of cut). Most workstations have considerable flexibility in the realm of crossfades, and are not limited to short durations. Indeed it is common now to use crossfades of many shapes and durations for different creative purposes, and this, coupled with the ability to preview edits and fine-tune their locations, has made it possible to put edits in places previously considered impossible.

The locations of edit points are kept in an edit decision list (EDL) which contains information about the segments and files to be replayed at each time, the in and the out points of each section, and details of the crossfade time and shape at each edit point. It may also contain additional information such as signal processing operations to be performed (gain changes, EQ, etc.)

5.4.4 Crossfading

Crossfading is similar to butt joining, except that it requires access to data from both incoming and outgoing files for the duration of the crossfade. The crossfade calculation involves simple signal processing, during which the values of outgoing samples are multiplied by gradually decreasing coefficients whilst the values of incoming samples are multiplied by gradually increasing coefficients. Time coincident samples of the two files are then added together to produce output samples. The duration and shape of the crossfade can be adjusted by altering the coefficients involved and the rate at which the process is executed.

Crossfades are either performed in real time, as the edit point passes, or pre-calculated and written to disk as a file. There are merits to both approaches. Real-time crossfades can be varied at any time and are simply stored as commands in the EDL, indicating the nature of the fade to be executed. The process is similar to that for the butt edit, except that as the edit point approaches samples from both incoming and outgoing segments

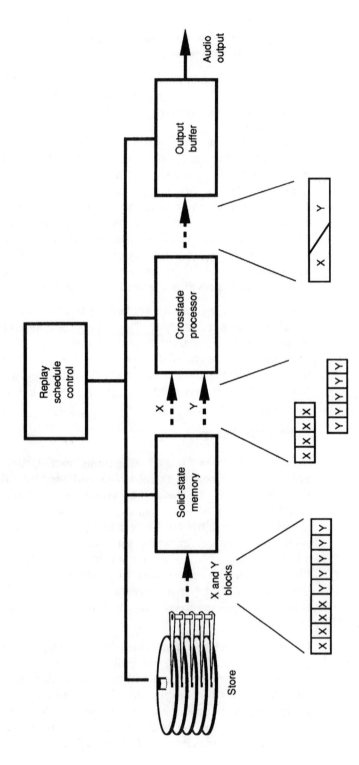

Figure 5.14 Conceptual diagram of the sequence of operations which occur during a crossfade. X and Y are the incoming and outgoing sound segments

are loaded into RAM in order that there is an overlap in time. During the crossfade it is necessary to continue to load samples from both incoming and outgoing segments into their respective areas of RAM, and for these to be routed to the crossfade processor, as shown in Figure 5.14. The resulting samples are then available for routing to the output. A consequence of this is that a temporary increase in disk activity occurs, because two streams of data rather than one are read during a crossfade. It is important, therefore, to have a disk drive and buffer size capable of handling the additional load present during real-time crossfades, which represents a doubling in the transfer rate required. Eight channel replay would effectively become sixteen channel replay for the duration of a crossfade edit on all eight channels, for example. An editing system may consequently be pushed close to its limits if asked to perform long real-time crossfades on multiple channels at the same time.

A common solution to this problem is for the crossfade to be calculated in non-real time when the edit point and crossfade duration is first determined by the user. This incurs a short delay while the system works out the sums, after which a new sound file is stored which is simply the crossfade period and nothing else. Replay of the edit is then a more simple matter, which involves playing the outgoing segment up to the beginning of the crossfade, then the crossfade file, then the incoming segment from after the crossfade, as shown in Figure 5.15. Load on the disk drive is therefore no higher than normal. This approach has advantages because it makes any number and length of crossfade possible on any combination of tracks, with the sure knowledge that they can be replayed. The slight disadvantage is the need for the system to write a new crossfade file every time the edit is altered, and the disk space taken up by the crossfade files (although this is normally quite small).

The shape of the crossfade is often able to be changed to suit a different operational purpose. Standard linear fades (those

Figure 5.15 Replay of a precalculated crossfade file at an edit point between files X and Y

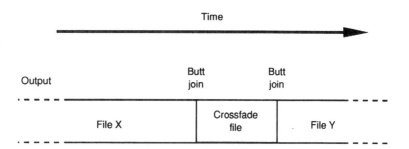

Figure 5.16 Summation of levels at a crossfade. (a) A linear crossfade can result in a level drop if the incoming and outgoing material are non-coherent. (b) An exponential fade, or other similar laws, can help to make the level more constant across the edit

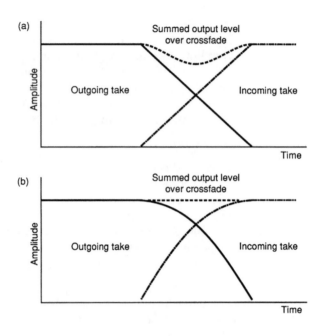

where the gain changes uniformly with time) are not always the most suitable for music editing, especially when the crossfade is longer than about ten milliseconds.[3] The result may be a momentary drop in the resulting level in the centre of the crossfade which is due to the way in which the sound levels from the two files add together. If there is a random phase difference between the signals, as there will often be in music, the rise in level resulting from adding the two signals will normally be around 3 dB, but the linear crossfade is 6 dB down in its centre resulting in an overall level drop of around 3 dB (see Figure 5.16). Exponential crossfades and other such shapes may be more suitable for these purposes, because they have a smaller level drop in the centre. It may even be possible to design one's own crossfade laws. Figure 5.17 shows the crossfade editing controls from a widely used system by Sonic Solutions. It is possible to alter the offset of the start and end of the fade from the actual edit point as well, and to have a faster fade up than fade down.

Many systems also allow automated gain changes to be introduced as well as fades, so that level differences across edit points may be corrected. Figure 5.18 shows a crossfade profile which has a higher level after the edit point than before it, and different slopes for the in and out fades. A lot of the difficulties which editors encounter in making edits work can be solved using a combination of these facilities.

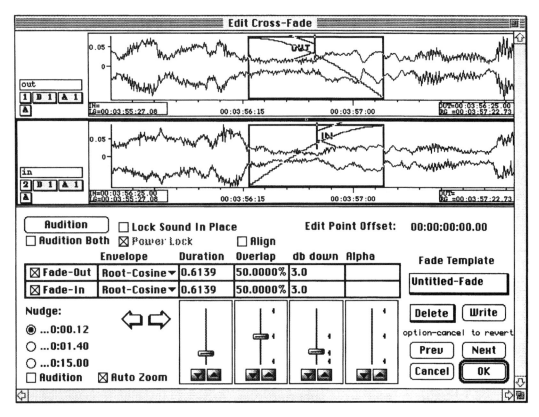

Figure 5.17 Crossfade manipulation controls from the Sonic Solutions digital editor, showing the various parameters which may be altered

Figure 5.18 The system may allow the user to program a gain profile around an edit point, defining the starting gain (A), the fade-down time (B), the fade-up time (D), the point below unity at which the two files cross over (C), and the final gain (E)

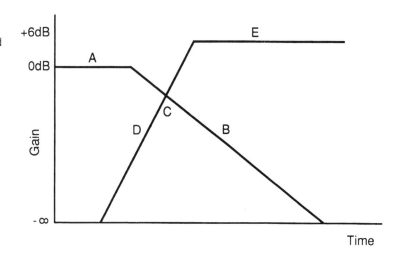

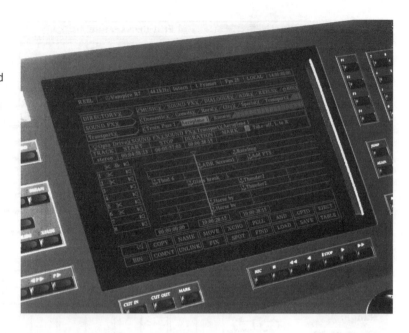

5.4.5 Editing modes

During the editing process the operator will load appropriate
sound files and audition them, both on their own and in a
sequence with other files. The exact method of assembling the
edited sequence depends very much on the user interface, but it
is common to present the user with a visual analogy of moving
tape, allowing him to 'cut-and-splice' or 'copy and paste' files
into appropriate locations along the virtual tape. These files are
then played out at the timecode locations corresponding to their
virtual position on this 'tape' (an example is shown in Figure
5.19). It is also quite common to display a representation of the
audio waveform which allows the editor to see as well as hear
the signal around the edit point (see Figure 5.20).

In the editing of music using digital tape systems it is common
to assemble an edited master from the beginning, copying takes
from source tapes in sequence onto the master. An example of
typical procedure will serve to illustrate the point. Starting at the
beginning of the piece of music the first take will be copied to
the master tape until a short time after the first edit is to be
performed. The editor will then locate the edit point on the
master tape (the outgoing take) by playing up to the approxi-
mate point and marking it, followed by fine trimming of this
point, either by nudging it in small time increments, or by the
simulation of analog 'reel-rocking' using a wheel which plays
out digital audio from memory at variable speed around the edit

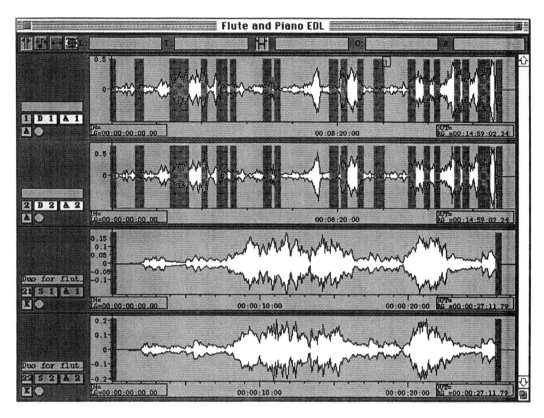

Figure 5.20 A waveform display is also commonly used to aid the editor in locating edit points. (Sonic Solutions editor)

point. The edit point is then confirmed and the same procedure is performed on the source take to be joined in at this point (the incoming take). This edit is now auditioned, with a crossfade between outgoing and incoming material at the edit point, after which any further trimming is performed before the edit is committed to the master tape by dropping it into record mode at the appropriate time.

In non-linear systems this approach is often simulated, allowing the user to roughly locate an edit point by playing the virtual tape followed by a fine trim using simulated reel-rocking. Sound files and segments are treated as the equivalent of the 'takes' in the above example, and the system notes the points in each segment at which one is to cease and another is to begin playing, with whatever overlap has been specified for crossfading.

Non-linear systems also make it possible to insert or change sections in the middle of a finished session. To take an example, assume that an edited opera has been completed and that the producer now wishes to change a take somewhere in the middle (see Figure 5.21). With tape-based copy editing it would be

159

Figure 5.21 Replacing a take in the middle of an edited programme. (a) Tape based copy editing results in a gap of fixed size, which may not match the new take length. (b) Non-linear editing allows the gap size to be adjusted to match the new take

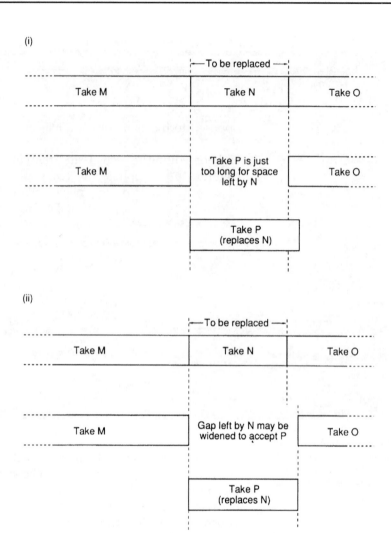

impossible to insert the replacement take without recopying all the material after that take, because the length of the new take would be unlikely to match the duration of the old one exactly. In non-linear editors it is possible simply to shuffle all of the following material along slightly to accommodate the new take, this being only a matter of changing the EDL rather than modifying the stored music in any way. The files are then simply played out at slightly different times than in the first version of the edit.

It is normal also to allow edited segments to be fixed in time if desired, so that they are not shuffled forwards or backwards when other segments are inserted. This 'anchoring' of segments is often used in picture dubbing when certain sound effects and dialogue have to remain in certain fixed locations with respect to the picture. This was illustrated in Figure 5.19, where little anchor symbols are displayed in each segment to show that their locations are fixed.

Disposal editing is also accommodated easily. This type of editing involves removing unwanted sections of sound, as opposed to the assembly of a master tape from stored sections. It is often used for removing unwanted noises from recordings, and for editing speech. Using the cut and paste approach so common to word processors it is possible to play up to the start of each segment to be removed, then stop the replay, finely search the start of the segment and 'cut' to the end of the segment by selecting a portion of the virtual tape to be removed. On replay, the system will simply skip the portion of the file which contains the unwanted sound, probably introducing a crossfade over the join, as shown in Figure 5.22. This is achieved by causing a jump in the file pointer which determines which block and sample of the sound file is to be read next. Again RAM buffering is used to iron out the discontinuity arising from the jump in disk reading position. The original sound file is not affected in any way, only the way it is replayed.

Figure 5.22 Unwanted material can easily be skipped on replay, as shown here

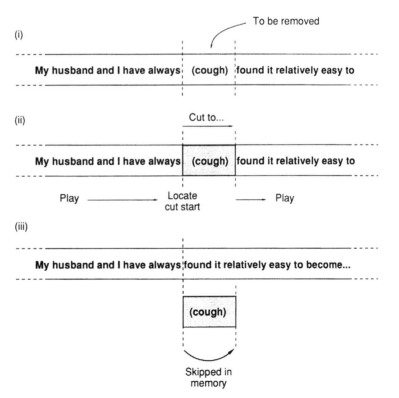

5.4.6 Simulation of 'reel-rocking'

It is common to simulate the effect of reel-rocking in non-linear editors, providing the user with the sonic impression that reels of analog tape are being 'rocked' back and forth as they are in analog tape editing when fine-searching edit points. Editors are used to the sound of tape moving in this way, and are skilled at locating edit points when listening to such a sound.

The simulation of variable speed replay in both directions (forwards and backwards) is usually controlled by a wheel or sideways movement of a mouse which moves the 'tape' in either direction around the current play location. The magnitude and direction of this movement is used to control the rate at which samples are read from the disk file, via the buffer, and this replaces the fixed sampling rate clock as the controller of the replay rate. Systems differ very greatly as to the sound quality achieved in this mode, because it is in fact quite a difficult task to provide convincing simulation. So poor have been the majority of attempts that many editors do not use the feature on non-linear editors, preferring to judge edit points accurately 'on the fly', followed by trimming or nudging them either way if they are not successful the first time. Good simulation requires very fast, responsive action and an ergonomically suitable control. A mouse is very unsuitable for the purpose. It also requires a certain amount of DSP to filter the signal correctly, in order to avoid the aliasing which can be caused by varying the sampling rate.

5.5 Timing and synchronisation

There are many cases in which it is necessary to ensure that the recording and replay of digital audio by a workstation are time-synchronised to an external reference of some sort. Under the heading of synchronisation comes the subject of locking both recording and replay to a source of SMPTE/EBU timecode or MIDI TimeCode (MTC), as well as locking to an external sampling rate clock, video sync reference or digital audio sync reference. This is needed when the workstation is to be integrated with other audio and video equipment, and when operating in an all-digital environment where the sampling frequencies of interconnected systems must be the same. The alternative, when the workstation is operating in isolation, is for all operations to be performed with relation to an internal timing reference locked to the prevailing audio sampling frequency.

It is normal for audio samples to be recorded and reproduced at one of the standard sampling frequencies described in Chapter 2, and there is therefore a need for a carefully controlled interrelationship between any external timing reference and the audio sampling frequency. It is the timing reference that has overall control of the rate at which data is read from and written to disk, since something has to ensure that neither too few nor too many samples are transferred per second, otherwise buffers would either overflow or empty.

5.5.1 Requirements for synchronisation

The synchronisation of a workstation requires that replay or recording speeds are kept in step with an external timing reference, and that there is no long-term drift between this external reference and the passage of time in the replayed audio signal. When a lock is required to an external reference there is the possibility that this reference may drift in speed, may have timing jitter, and may 'jump' in time (if it is a 'real-time' reference such as timecode). Such situations would require that the replay speed and sampling rate of the workstation be adjusted regularly and continuously to follow any variations in the timing reference, and this can give rise to audible artefacts due to sample jitter (see section 2.4), or to the variation of the output sampling rate outside the tolerances acceptable by any other digitally interfaced device in the system (see section 6.4).

In order to achieve synchronisation with an external source without these problems it is necessary to implement both smoothing of the variations in the timing reference, and perhaps variable rate sampling frequency conversion to ensure a constant rate output no matter what occurs internally. In the case of analog interfacing, some of the problems associated with variations in the output sampling rate may not be so important, although this depends on the speed at which they occur.

5.5.2 Timecode synchronisation

The most common synchronisation requirement in non-linear recording is for replay to be locked to a source of SMPTE/EBU timecode, because this is used universally as a timing reference in audio and video tape recording, as well as in other operations. Details of this timecode are given in section 7.6 for those who are not yet familiar with it. A number of desktop workstations that have MIDI features lock to MIDI TimeCode (MTC), which is a representation of SMPTE/EBU timecode in the form of MIDI messages, normally provided by an external MIDI and timecode interface, as discussed further in section 8.10.

A form of timecode is used extensively as an internal reference within the tapeless system, because it provides a measure of time against which the replay of sound files can be executed. There may not be a direct integer relationship between sector address and timecode increment, as there are a number of timecode frame rates, and the possibility for a variety of sampling rates. For example, at 48 kHz sampling rate, sixteen bits per sample, one sector of 512 bytes corresponds to a time period of 5.33 ms, but at 44.1 kHz sampling rate it corresponds to 5.8 ms. It is clear then that, although it might be desirable for there to be a fixed relationship between disk sectors and timecode, any relationship will only hold true for one particular sampling rate and timecode frame rate.

One approach to this problem (and there are many) is to use a different 'timecode' internally to the workstation, which relates to the digital audio sample structure, and which can thereafter be related to the SMPTE/EBU timecode by suitable offsets and 'gearbox' ratios. Such a timecode might take the form of a 'sample address code' in which time is measured by the number of sample periods elapsed since a nominal midnight. All internal references within the workstation could be made using this sample address code, such as segment start and end times, edit points, and so forth, and this would allow sample-accurate location of material. For display and synchronisation purposes the sample address could be converted into whatever timing standard was appropriate for the application, assuming that the frame rate and sampling rate were known.

From the external timing reference a sample address could be derived which (in the case of SMPTE/EBU timecode) would be a conversion from hours, minutes, seconds and frames to a sample address, this being achieved using a software 'gearbox' which relates the frame rate to the sampling rate. Offsets could then be introduced if necessary.

In workstations used extensively with video and film it is common for sound file headers or resource forks (see section 6.1) to contain information such as the SMPTE/EBU start time and frame rate when the file was recorded, and possibly the musical bar and beat number for synchronisation with MIDI sequencers. This allows files to be synced up to their original time locations in a programme if that is required, or to be offset by a fixed amount of time.

5.5.3 True synchronisation or simple triggering?

It is important to know what kind of synchronisation is used in your workstation. One of the factors which must be considered is whether external timecode is simply used as a timing reference against which sound file replay is triggered, or whether the system continues to slave to external timecode for the duration of replay. In some cases these modes are switchable because they both have their uses. In the first case replay is simply 'fired off' when a particular timecode is registered, and in such a mode no long-term relationship is maintained between the timecode and the replayed audio. This may be satisfactory for some basic operations, but is likely to result in a gradual drift between audio replay and the external reference if files longer than a few seconds are involved. It may be useful though, because replay remains locked to the workstation's internal clock reference, which may be more stable than external references, potentially leading to higher audio quality from the system's convertors. Some cheaper systems do not 'clean up' external clock signals very well before using them as the sample clock for D/A conversion, and this can seriously affect audio quality, as described in Chapter 2.

In the second mode a continuous relationship is set up between timecode and audio replay, such that long-term lock is achieved and no drift is encountered. This is more difficult to achieve because it involves the continual comparison of timecode to the system's internal timing references, and requires that the system follows any drift or jump in the timecode. The rate at which samples are transferred to and from the disk memory buffers is then controlled by the external reference. Jitter in the external timecode is very common, especially if this timecode derives from a video tape recorder, and this should be minimised in any sample clock signals derived from the external reference. This is normally achieved by the use of a high quality phase-locked loop, often in two stages.[4] Wow and flutter in the external timecode can be smoothed out using suitable time constants in the software which converts timecode to sample address codes, such that short-term changes in speed are not always reflected in the audio output but longer-term drifts are.

Sample frequency conversion may be employed at the digital audio outputs of a workstation to ensure that changes in the internal sample rate of the workstation caused by synchronisation action are not reflected in the output sampling rate. This may be required if the workstation is to be interfaced digitally to other equipment in an all-digital studio.

5.5.4 Synchronisation to external digital audio, film or video references

In all-digital systems it is necessary for there to be a fixed sampling frequency, to which all devices in the system lock. This is so that digital audio from one device can be transferred directly to others without conversion or loss of quality. In systems involving video it is often necessary for the digital audio sampling rate to be locked to the video frame rate, and for timecode to be locked to this as well. A similar process is involved here to that involved in timecode synchronisation, except that here the reference signal is likely to be a 'house sync' signal which does not necessarily carry time-of-day information. It would be used to lock the internal sampling frequency clock of the workstation. Other sync signals could include tachometer or control track pulses from tape machines or frame rate pulses from film equipment. If a system is to be able to resolve to any or all of these, as well as to timecode and digital audio inputs, a very versatile 'gearbox' will be required to perform the relevant multiplications and divisions of synchronisation signals at different rates, such that they can be used to derive the internal sampling rate clock of the system. A stable voltage-controlled oscillator (VCO) and phase-locked loop are commonly used for this purpose.

Figure 5.23 illustrates a possible conceptual diagram of synchronised operation, with a variety of references and a constant sampling rate output. The topic of digital audio and video signal synchronisation is covered in more detail in section 6.5.

Figure 5.23 Conceptual diagram of replay synchronised to one of a number of timing sources. Blocks of data are fetched from disk at a rate determined by the current sampling clock

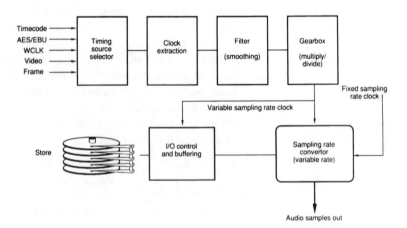

5.6 Variable speed operation

In analog tape recording it is common to replay audio at speeds other than the nominal play speed. The most common situations are those in which either (a) the tape speed is changed by a small percentage above or below the nominal to alter the pitch of the sound, (b) the reels are rocked backwards and forwards by hand to locate an edit point, or (c) the tape is spooled at fixed multiples of play speed. In (a) one is dealing with changes of speed around the standard play speed, whereas in (b) one is dealing with speeds ranging upwards from zero, in two directions. Furthermore, (b) may involve considerable acceleration of the tape reels, and thus fast changes in the speed of the tape. Case (c) involves spooling forwards or backwards at higher than normal play speed.

In order to produce variable speed replay in a workstation one must vary the rate at which samples are read out of the disk/buffer combination. This has the effect of changing the audio sampling rate, and, as described in Chapter 2, it is possible to create aliased products by lowering the sampling rate too far. The higher the nominal sampling rate, the further downwards it may be moved in varispeed modes before aliasing begins to occur. Digital low-pass filtering may be employed to remove components above half the sampling frequency, which will act to prevent aliasing, and this may be combined with sample rate conversion to ensure a constant output rate whilst the replay rate changes.

The limits on the amount of varispeed allowed in a disk-based system depend on a number of factors, particularly the maximum rate at which data may be read from the store. For example, a storage device which is already used close to its transfer rate limits will not be capable of supplying data at a rate much higher than that required for normal speed replay.

The distortion requirements of reel-rocking simulation are perhaps somewhat different to those needed for small percentages of varispeed around play speed, as it is more important in the latter case that the highest audio quality is preserved, whereas in the former one is concerned more with the accurate location of an edit point and may not be so concerned with audio quality. Furthermore, the means of control for reel-rocking simulation is likely to be a large wheel which can be moved in either direction at high speed, whereas the varispeed control is likely to be a small vernier control. Digital filters have been developed which offer high audio quality over a very wide range of input sampling rates, whilst maintaining a constant output sampling rate, and thus similar DSP techniques may be employed for both applications.

For the simulation of audible high-speed spooling, samples may be read from the disk in the normal way up to its maximum transfer rate, but this may only allow for spooling at perhaps two or three times play speed (depending on the speed of the disk and the number of channels it serves). In many cases this may be acceptable, and longer distances may be covered by 'go to' commands. If audible replay at higher spooling speeds is required then it is possible that so-called 'spooling files' will have to be written at the time of recording, these being resolution-reduced versions of the main files. For high speed replay, the spooling file would be played out at a higher rate than that at which it was recorded, simulating the desired effect.

5.7 DSP in workstations

The majority of audio workstations have some signal processing capabilities in addition to basic recording, replay and editing functions. These can range from the most basic level controls to advanced features such as time compression and noise reduction. An introduction to the principles of DSP was given in Chapter 2.

5.7.1 Mixing

The most common DSP operation included in workstations is digital mixing. This allows for complete productions to be executed on the workstation without needing any additional audio equipment, with a final stereo or multichannel output being made available for dumping to the appropriate master format. External inputs and outputs (either analog or digital) can often be brought into this mix as well, by using the audio interfaces of the workstation to route signals to and from external sources such as effects units. Alternatively one can implement all the effects one needs using on-board signal processing and internal routing of audio signals.

The mixing interface may either be an on-screen collection of faders and other controls (Figure 5.24(a)) or it may be an external physical control panel looking very much like a conventional mixer (Figure 5.24(b)). The former is normally cheaper but quite a lot more difficult to use. Software control usually allows the mixer to be automated, so that fader movements and other settings can be stored in a dynamic fashion (control positions can be stored on a continually updated basis). It may be possible to display the gain profile of a track graphically so that it can be edited by using a mouse to drag points up or down, or add further points at which the level will change, as shown in Figure 5.25. Panning can be controlled in the same way.

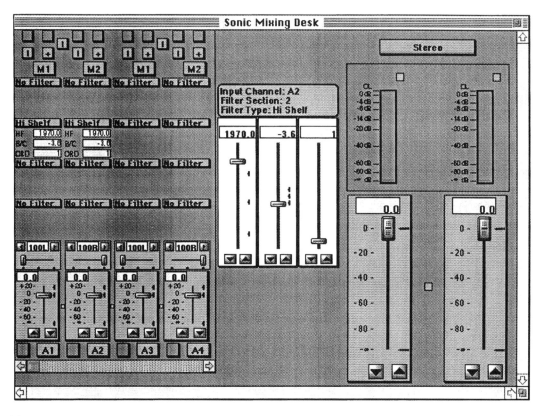

(a)

Figure 5.24 Mixer control interfaces. (a) On-screen mixer from Sonic Solutions. (b) A dedicated mix controller (DAR SoundStation Gold)

(b)

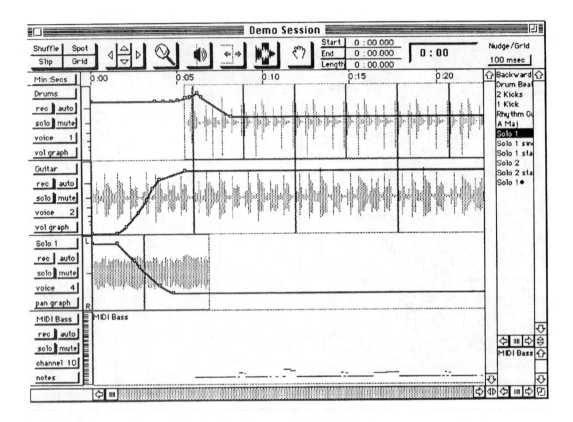

Figure 5.25 Audio processing options such as volume and pan can be dynamically automated and displayed in the form of a gain 'contour' as shown in this example from ProTools. Track 1 shows a volume contour, and Track 2 shows a pan contour

5.7.2 'Segment-based' processing

When applying effects or other processing such as gain modification or EQ to a production, one generally applies them to a particular audio segment on a particular track. Problems can arise if that segment is subsequently moved, either in time or to another track, because the processing may remain fixed to the original location. DAR coined the term 'segment-based processing' to mean processing which could be attached firmly to the segment concerned, following it to any new location in the production. If you copy the segment, the processing goes with it. Processing from one segment can also be copied to another without changing the audio in any other way.

5.7.3 Internal digital audio routing and effects

Workstations which aim to incorporate expandable signal processing resources usually make use of a high speed bus to carry audio between the various parts of the system. Digidesign's TDM bus is used here as an example of such an approach, because it shows how signal processing can be built

Figure 5.26 Signal processing and I/O hardware are all connected to a common high speed bus in Digidesign's TDM system, allowing third party hardware to be incorporated

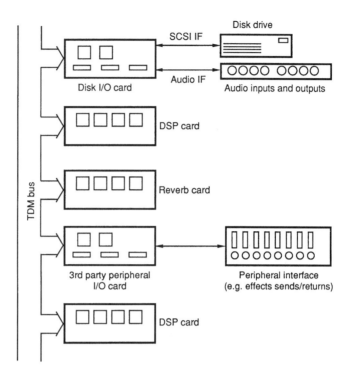

up in a modular fashion, even allowing for hardware and software from other manufacturers to be incorporated within the structure.

The TDM bus is basically a time-division multiplexed bus which can carry audio data for up to 256 channels at 24 bit resolution, and which can be used to interconnect processor cards so that audio can be routed and transferred between them. An example of such an approach is shown in Figure 5.26. The TDM bus is totally separate from the computer's expansion bus (which would never be fast enough for such a purpose). As can be seen, it is possible to include cards from third parties which allow the user to choose the most suitable devices for a particular purpose. It would be possible to have audio I/O's via a preferred company's convertors, for example, and reverberation from another source. Using a universal software driver called DAE (Digidesign Audio Engine), third party software applications can hook into the system and appear as routing sources and destinations in the on-screen mixer.

The company allows users to buy DSP capacity as required, each DSP card being linked using the TDM bus, on which can be run a number of third-party 'plug-in' applications that are integrated with the standard user interface.

5.7.4 Time compression and expansion

Digital signal processing makes possible the time expansion or compression of audio passages without changing the pitch. The sound material to be treated is first analysed, followed by intelligent contraction or expansion in the time domain by editing the waveform at strategic points so that a new file is created which is either longer or shorter than the original passage. If then played at the nominal sampling rate, this file will appear to be the same passage, but at a faster or slower speed than before, with the original pitch. Done well the technique can produce reasonable sounding audio, but there are some fairly basic algorithms in existence which produce audible 'glitches' in sustained musical notes. As with most signal processing, the more expensive systems tend to produce better results.

With music, the technique is generally only successful over a range of around 10 per cent either side of the nominal speed, otherwise the sound quality begins to deteriorate, but with speech it may be possible to time-compress material by up to 50 per cent, and this may be used to solve problems such as that of needing to fit a commentary over a particular length of picture during dubbing. It is possible that the processing time required to calculate the time-modified file may be considerable, and for this reason it may not always be possible to perform the operation 'on-the-fly' in real time.

Jeff Bloom developed an algorithm, implemented as 'Wordfit'[5] in DAR's products, which was able to analyse a guide audio track, such as a location dialogue track, and automatically 'fit' an overdubbed track to it so that it matched the time varying nature of the original track. This proved particularly useful in post-dubbing dialogue because it removed the need for dubbing artists to achieve correct lip sync, thereby speeding up the process. It has also proved workable in a number of cases with overdubs in a foreign language.

Wordfit works by performing a regular spectrum analysis of the two audio signals (every 10 ms) and creating a 'time-warping' path which represents the timing modifications that need to be made to the overdubbed track in order for its key elements to match those found in the guide track. Silences in the speech are also identified, since it is easy to condense or extend the gaps between words. The overdubbed audio is then edited automatically at points in the passage where a timing alteration is required, by analysing the periodic nature of the wave and performing a pitch-synchronous edit which either adds or removes about 10 ms of the waveform without changing the pitch.

5.7.5 Digital noise reduction

A number of manufacturers now offer DSP hardware and software to perform noise reduction or removal on programme material. Unlike most analog tape based noise reduction systems, which require encoding and decoding of the audio signal, these digital systems are capable of isolating the offending noise from the signal and then subtracting it, or using adaptive digital filters to 'tune out' pitched noises such as hum and buzz. In the case of broadband noise it is normal for the system to use a 'silent' section of the programme to analyse the noise spectrum and level, from which it derives a template of the noise signal. This template is then used as the basis for subtraction during programme. For other types of noise it is possible to choose a variety of types of digital filter whose characteristics can be finely tuned to those of the unwanted noise, requiring considerable operator skill in many cases to remove noises without unduly affecting the quality of the remaining programme. For example, one might construct a filter to remove the hum from a track. Hum usually has a number of harmonics at multiples of the power line frequency, and digital filters can be created which have notches at all the appropriate frequencies so as to remove not only the fundamental but also the annoying overtones of the hum.

The process is not a complementary encode/decode process, such as Dolby noise reduction, and there can sometimes be noticeable side effects of digital single-ended noise reduction systems. One has to trade off the degree of noise reduction with the side effects, and it is rare for good quality results to be obtained without quite careful adjustment of the parameters involved.

References

1   Anderson, D. (1993) 'High speed networking for professional digital audio'. AES UK Digital Audio Interchange Conference, 18–19 May, London, p. 62. Audio Engineering Society.
2   Abbot, C. (1984) 'Efficient editing of digital sound on disk'. Journal of the Audio Engineering Society, 32, 6, pp. 394–402.
3   Turner, B. (1990) 'Music editing on hard disk systems'. AES UK Hard Disk Recording Conference, 16–17 May, London, pp. 51–56. Audio Engineering Society.
4   Parker, M. (1990) 'Synchronisation of disk based systems'. AES UK Hard Disk Recording Conference, 16–17 May, London, pp. 72–81. Audio Engineering Society.
5   Bloom, P. *et al.* (1987) 'A digital signal processing system for automatic dialogue post synchronisation'. Presented at the 83rd AES Convention, New York, 16–19 October, preprint 2546 (K-7).

6 File formats and data interchange

This chapter is all about the formats in which audio and related data are stored and exchanged between systems. It explains a number of issues relating to audio file formats used commonly in workstations, introduces the Open Media Framework Interchange, describes commonly used real-time digital audio interfaces, discusses sample rate synchronisation and introduces the use of networks for transferring digital audio. It also includes details of some of the most commonly encountered formats for Compact Disc pre-mastering.

6.1 Audio file formats

6.1.1 Background

There are almost as many file formats for audio as there are days in the year. For a long time the specific file storage strategy used for disk-based digital audio was the key to success in digital workstation design, because disk drives were relatively slow and needed clever strategies to ensure that they were capable of handling a sufficiently large number of audio channels. Manufacturers also worked very much in isolation, and the size of the market was relatively small, leading to virtually every workstation type using a different file format for audio and edit list information.

There are still advantages in the use of filing structures specially designed for real-time applications such as audio and video editing, because one will always obtain better performance from a disk drive in this way, but the need is not as great as it used to be. Interchange is becoming at least as important, if not more important than ultimate transfer speed, and the majority of hard disk drives available today are capable of replaying many channels of audio in real time without a particularly fancy storage strategy. Indeed a number of desktop systems simply use the native filing structure of the host computer, as introduced in Chapter 4. As the use of workstations grows, the need for files to be transferred between systems also grows, and either by international standardisation or by sheer force of market dominance certain file formats are likely to become the accepted means by which data are exchanged. This is not to say that we will only be left with one or two formats, but that systems will have to be able to read and write files in the common formats if users are to be able to send and receive work from others.

It is not proposed to attempt to describe all of the file formats in existence, since that would be a relatively pointless exercise and would not make for interesting reading. It is none the less useful to have a look at some examples taken from the most commonly encountered file formats, particularly those used for high quality audio by desktop and multimedia systems, since these are amongst the most widely used in the world and are often handled by audio workstations even if not in their native format. It is not proposed to investigate the large number of specialised file formats developed principally for computer music on various platforms, nor the files used for internal sounds and games on many computers. Following this the Open Media Framework Interchange (OMFI) standard will be described, which is the closest thing that exists at the moment to a universal interchange standard for digital audio, video and edit list information.

It is recommended that the first part of this chapter is read in conjunction with the sections of Chapter 4 dealing with the structure and formatting of computer mass storage media, because this is another vital element in the understanding of media interchange between systems.

6.1.2 Audio file formats in general

A data file is simply a series of data bytes formed into blocks and stored either contiguously or in fragmented form. In a sense files themselves are independent of the operating system and filing structure of the host computer, because a file can be transferred to another platform and still exist as an identical series of data blocks. It is the filing *system* which is often the platform- or operating system-dependent entity. There are some features of data files which relate directly to the operating system and filing system which created them, these being fairly fundamental features, but they do not normally prevent such files being translated by other platforms.

For example, there are two approaches to byte ordering: the so-called little-endian order in which the most significant byte comes first, and the big-endian format in which the least significant byte comes first. These relate to the byte ordering used in data processing by the two most common microprocessor families, and thereby to the two most common operating systems used in desktop audio workstations. Motorola 680x0-series processors, as used in the Apple Macintosh, deal in little-endian byte ordering, and Intel 80x86 processors, as used in MS-DOS machines, deal in big-endian byte ordering. It is relatively easy to interpret files either way around, but it is necessary to know that there is a need to do so if one is writing software.

Secondly, there is the fact that Macintosh files may have two parts: a resource fork and a data fork, as shown in Figure 6.1. High level 'resources' are stored in the resource fork, (used in some audio files for storing information about the file, such as signal processing to be applied, display information and so forth) whilst the raw data content of the file is stored in the data fork (used in audio applications for audio sample data). The resource fork is not always there, but may be. This is not common to most other computer platforms which tend to store all the information in one part.

Some data files include a 'header', which is a number of bytes at the start of the file which contain information about the data

Figure 6.1 Macintosh files may have a resource and a data fork

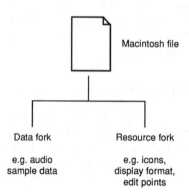

Figure 6.2 Many files include a header which comes before the main data, as shown in (a), whereas others may simply contain 'raw data', as shown in (b)

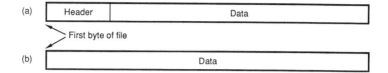

(a)

| Header | Data |

First byte of file

(b)

| Data |

that follows (see Figure 6.2). In audio systems this may be things like the sampling rate and resolution of the file. Audio replay would normally be started from immediately after the header. On the other hand, some files will be simply raw data, usually in cases where the format is fixed. ASCII text files are a well known example of raw data files, since they simply begin with the first character of the text.

The audio data in most common high quality audio formats is stored in two's complement encoded form (see section 1.3.2), and the majority of files are used for 16 bit data, thus employing exactly two bytes per audio sample. 8 bit files use one byte per sample. File formats requiring more than 16 bit resolution have two options: if they want to allow for resolution up to 24 bits then a simple 3 bytes per sample is appropriate, but this is wasteful of storage if 20 bit samples are the norm. In such a case some file formats share a byte between two samples in various ways, using it for four bits of one sample and four bits of the other.

6.1.3 Sound Designer I format

Sound Designer files originate with the Californian company Digidesign, manufacturer of probably the world's most widely used digital audio hardware for desktop computers. Many systems handle Sound Designer files because they are used for such purposes as the distribution of sound effects on CD-ROM, and for other short musical sample files. Detailed information about Digidesign file formats can be obtained if one wishes to become a third-party developer, and the company exercises no particular secrecy in the matter.[1]

The Sound Designer I format (SD I) is for mono sounds, and it is recommended for use in storing principally short sounds. It originated on the Macintosh, so numerical data are stored in little-endian byte order, and it has no resource fork. The data fork contains a header of 1336 bytes which is followed by the audio data bytes. The header contains information about how the sample should be displayed in Sound Designer editing software, including data describing vertical and horizontal

177

scaling, and it also contains details of 'loop points' for the file (these are principally for use with audio/MIDI sampling packages where portions of the sound are repeatedly cycled through while a key is held down, in order to sustain a note). The header also contains information on the sample rate, sample period, number of bits per sample, quantisation method (e.g. 'linear' or some other form, expressed as an ASCII string describing the method) and size of RAM buffer to be used.

The audio data is normally either 8 or 16 bit, and is always MSbyte followed by LSbyte of each sample.

6.1.4 Sound Designer II format

Sound Designer II is one of the most commonly used formats for audio workstations, and has greater flexibility than SD I. Again it originated as a Mac file and unlike SD I it has a separate resource fork. For this reason the data fork contains simply the audio data bytes, again in two's complement form and either 8 or 16 bits per sample. SD II files can contain audio samples for more than one channel, in which case the samples are interleaved, as shown in Figure 6.3, on a sample by sample basis (i.e. all the bytes for one channel sample followed by all the bytes for the next, etc.). It is unusual to find more than stereo data contained in SD II files, and it is recommended that multichannel recordings are made using separate files for each channel. Some multichannel applications, when opening stereo SD II files have first to split them into two mono files before they can be used, by deinterleaving the sample data. This requires that there is sufficient free disk space for the purpose.

Figure 6.3 Sound Designer II files allow samples for multiple audio channels to be interleaved. 4 channel, 16 bit example shown

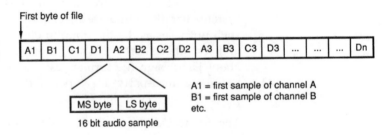

Since Mac resource forks are able to be written separately from their associated data forks, it is possible to update the descriptive information about the file separately from the audio data. This can save a lot of time (compared with single-fork files such as SD I) if the file is long and the audio data has not changed, because it saves rewriting all the audio data at the same time. Only three resources are mandatory, and others can be added by developers to customise the files for their own purposes. The mandatory ones are 'sample size' (number of bytes per sample), 'sample rate' and 'channels' (describing the number of audio channels in the file). Digidesign optionally adds other resources describing things like the timecode start point and frame rates originally associated with the file concerned, for use in post-production applications.

6.1.5 AIFF and AIFF-C formats

The AIFF format[2] is widely used as an audio interchange standard, because it conforms to the EA IFF 85 standard for interchange format files used for various other types of information such as graphical images.[3] AIFF is an Apple standard (little-endian) format for audio data, and as such is encountered widely on Macintosh-based audio workstations, as well as being one of the standard audio interchange methods included in OMFI (see section 6.2). It is claimed that AIFF is suitable as an everyday audio file format as well as an interchange format, and some systems do indeed use it in this way. It can store audio information at a number of resolutions and for any number of channels if required, and the related AIFF-C (file type 'AIFC') format allows also for compressed audio data. It consists only of a data fork, with no resource fork, making it easier to transport to other platforms.

All IFF-type files are made up of 'chunks' of data which are typically made up as shown in Figure 6.4. A chunk consists of a header and a number of data bytes to follow. The simplest AIFF files contain a 'common chunk', which is equivalent to the header data in other audio files, and a 'sound data' chunk containing the audio sample data. These are contained overall by a 'form' chunk, as shown in Figure 6.5. AIFC files must also contain a 'Version Chunk' before the common chunk to allow for future changes to AIFC.

Figure 6.4 General format of an IFF file chunk

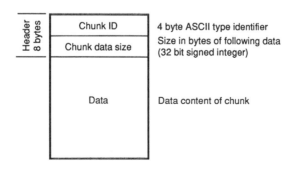

179

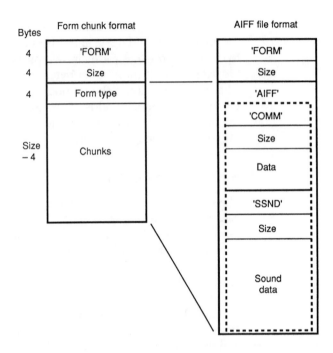

Figure 6.5 General format of an AIFF file

The common chunk header information describes the number of audio channels, the number of audio samples per channel in the following sound chunk, bits per sample (anything from 1 to 32), sample rate, compression type ID (AIFC only, a register is kept by Apple), and a string describing the compression type (again AIFC only). The sound data chunk consists of two's complement audio data preceded by the chunk header, the audio samples being stored as either 1, 2, 3 or 4 bytes, depending on the resolution, interleaved for multiple channels in the same way as for SD II files (see above). Samples requiring less than the full 8 bits of each byte should be left justified (shifted towards the MSB), with the unused LSBs set to zero.

Optional chunks may also be included within the overall container chunk, such as marker information, comments, looping points and other information for MIDI samplers, MIDI data (see section 8.3), AES channel status data (see section 6.4), text and application-specific data.

6.1.6 RIFF WAVE format

The RIFF WAVE format is the Microsoft equivalent of Apple's AIFF. It has a similar structure, again conforming to the IFF pattern, but with numbers stored in big-endian rather than little-endian form. It is used widely for sound file storage and inter-change on PC workstations, and for multimedia applications involving sound. Within WAVE files it is possible to include information about a number of cue points, and a playlist to indicate the order in which the cues are to be replayed.

6.1.7 MPEG audio file formats

The MusiFile format was developed by Digigram[4] and the CAR-Group to fill the need for a standard file format for MPEG audio (see Chapter 2). The MusiFile format consists of a header describing the nature of the file and a data area specific to the manufacturer, followed by data-reduced audio samples. It is also possible to store MPEG-compressed audio in AIFF-C or WAVE files, with the compression type noted in the appropriate header field, and it is proposed that MPEG audio files will be included within OMFI in future revisions, but no consensus has yet emerged. There are also MS-DOS file extensions used to denote MPEG audio files, notably .MPA (MPEG Audio) or .ABS (Audio Bit Stream).

6.1.8 Sampler files and formats

There are a large number of different MIDI-based audio samplers on the market, all of which use different storage formats for audio sample data on disk. The most common format is probably that used by Akai samplers such as the S-1000 and S-3000. Whilst these are not the same as any of the formats described above, there are various software utilities that will allow these sampler files to be read by audio workstation applications, translating them into one of the more common file formats. This allows the very large commercial libraries of samples, often distributed on CD-ROM, to be made available to audio workstations and not just to MIDI samplers.

It is also possible on certain multimedia workstations to trans-fer straight audio tracks, or sections thereof, from CD-Audio disks via a CD-ROM drive's SCSI interface to the computer's hard disk. There they are stored in one of the common inter-change formats such as WAVE or AIFF, which makes it easy to use them in audio editing applications. The ability to perform this operation depends very much on the driver software for the CD-ROM drive in question, and the availability of appropriate software for the hardware platform, because it is not possible in all cases.

6.1.9 Edit decision list (EDL) files

EDL formats are usually unique to the workstation on which they are used. That said, there is a widely used format for EDLs in the video world which is known as the CMX-compatible form. CMX is a well-known manufacturer of video editing equipment and most editing systems will read CMX EDLs for the sake of compatibility. These can be used for basic audio purposes, and indeed a number of workstations can read CMX EDL files for the purpose of auto-conforming audio edits to video edits performed on a separate system. The CMX list defines the cut points between source material and the various transition effects at joins, and it can be translated reasonably well for the purpose of defining audio cut points and their timecode locations, using SMPTE/EBU form.

The OMFI structure also contains a format for interchanging edit list data, as described below. Software can also be obtained for audio and video workstations which translates EDLs between a number of different standards to make interchange easier.

6.2 The Open Media Framework Interchange (OMFI)

OMFI was introduced in 1994 by Avid Technology, an American company specialising in desktop audio and video post-production systems (now merged with Digidesign). It was an attempt to define a common standard for the interchange of audio, video, edit list and other multimedia information between workstations running on different platforms, and it represents the best attempt so far at establishing a standard in this field. It is in effect a publicly available format, and Avid does not charge licensing fees of any kind, OMFI being a means of trying to encourage greater growth in this field of the industry as a whole. A number of other manufacturers have signed up to support OMF, and are jointly working on its development. Demonstrations have taken place of different workstations sharing common files and edit lists, which appear to have been successful. Avid makes available an OMF Interchange Toolkit at moderate cost for developers who want to build OMF compatibility into their products.

OMFI files contain two types of information: 'compositions' and 'sources'. Compositions specify how the various sources are to be assembled in order to play the finished product. Source data (audio, video, or other multimedia files) may be stored either in separate files, referenced by the OMF file, or within the OMF container structure. The container structure is similar to the IFF model described above (indeed Avid originally started to use IFF), in that it contains a number of self-describing parts, and is called Bento[5] (an Apple development). Each part of the OMFI file is complete in itself and can be handled independently of the other parts – indeed applications do not need to be able to deal with every component of an OMFI file – allowing different byte ordering for different parts if required. There is something of a danger in this 'all things to all men' approach, because systems may claim OMFI compatibility yet still not be able to deal with some of the data objects contained within the file.

The OMFI 1.0 specification[6] is quite lengthy and deals with descriptions of the various types of information that can be contained and the methods of containment. It also contains details of compositions and the ways in which edit timing data is managed. As far as the audio user is concerned, the 1.0 version specifies that the common audio formats to be used are the uncompressed versions of either the AIFF format or the WAVE format (see above), depending on the intended hardware platform. It also allows for the possibility that manufacturers might want to specify 'private' interchange formats of their own. Most of the version 1.0 document refers to video operations, so cuts and effects are all described in video terms and there is very little that refers to audio crossfades at edit points. Version 1.0 of OMFI is very video oriented, and specifies no more for audio than the two common formats for the audio data files and a means of specifiying edit points and basic crossfade durations (but not the shape), assuming that a video dissolve could mean the same as an audio crossfade. This is indeed a start, and Avid is proposing that the next version of OMFI will also contain a means of transferring audio volume, pan and EQ information.

It has often been said that what people really want in terms of interchange is the ability to take a disk or tape from one system and use it directly with another, but this is increasingly unrealistic because of the wide variety and rapid evolution of storage media and filing systems. Although people may use optical disks or tape cartridges as a means of transferring OMFI files between workstations from different manufacturers, the standard does not say anything about this, neither does it say anything about the filing system or hardware platform associated with the files. Those working with OMFI at Avid claim to have investigated a platform independent filing structure known as TAR which might be recommended as a possible move towards being able to specify more about a standard filing structure for OMFI data. TAR drivers could then be used on any hardware platform needing to mount volumes containing OMFI data, no matter what filing system was native to the platform concerned.

OMFI audio files can be used as the native format for a workstation's audio storage. This might not be an optimal solution from a performance point of view, but the standard has apparently been designed for this to be an option.

6.3 CD pre-mastering formats

The original tape format for submitting CD masters to pressing plants was Sony's audio-dedicated PCM 1610/1630 format on U-matic video tape. This is now 'old technology' and is rapidly being replaced by alternatives based on more recent data storage media and file storage protocols. These include the PMCD (pre-master CD) and Sony's MasterDisc format. A promising format, and likely to become a *de facto* standard, is the Disk Description Protocol (DDP) developed by Doug Carson and Associates.[7] Version 1 of the DDP laid down the basic data structure but said little about higher level issues involved in interchange, making it more than a little complicated for manufacturers to ensure that DDP masters from one system would be readable on another, but it is expected that Version 2, due soon, will address some of these issues.

DDP is a protocol for describing the contents of a CD, which is not medium specific. That said, at the time of writing it has become the norm to interchange DDP data on 8 mm Exabyte data cartridges. It can be used for interchanging the data for a number of different CD formats, such as CD-ROM, CD-DA, CD-I and CD-ROM-XA, and the protocol is really extremely simple. It consists of a number of 'streams' of data, each of which carries different information to describe the contents of the disk. These streams may be either a series of packets of data transferred over a network, files on a disk or tape, or raw blocks of data independent of any filing system. The DDP protocol simply maps its data into whatever block or packet size is used by the medium concerned, provided that the block or packet size is at least 128 bytes. Either a standard computer filing structure can be used, in which case each stream is contained within a named file, or the storage medium is used 'raw' with each stream starting at a designated sector or block address.

Manufacturers have discovered that the block length of DDP masters on Exabyte tapes does not remain constant between the different streams described below, usually being the equivalent of 4 CD sectors in the main stream, but smaller for other information. The ANSI tape labelling specification, described at the end of Chapter 4, is used to label the Exabyte tapes used for DDP transfers. This allows the names and locations of the various streams to be identified.

The principal streams included in a DDP transfer are as follows:

1 *DDP ID stream or 'DDPID' file.* 128 bytes long, describing the type and level of DDP information, various 'vital statistics' about the other DDP files and their location on the medium (in the case of physically addressed media), and a user text field (not transferred to the CD).

2 *DDP Map stream or 'DDPMS' file.* This is a stream of 128 byte data packets which together give a map of the CD contents, showing what types of CD data are to be recorded in each part of the CD, how long the streams are, what types of subcode are included, and so forth. Pointers are included to the relevant text, subcode and main streams (or files) for each part of the CD.

3 *Text stream.* An optional stream containing text to describe the titling information for volumes, tracks or index points (not currently stored in CD formats), or for other text comments. If stored as a file, its name is indicated in the appropriate map packet.

4 *Subcode stream.* Optionally contains information about the subcode data to be included within a part of the disk, particularly for CD-DA. If stored as a file, its name is indicated in the appropriate map packet. Apparently subcode blocks on DDP Exabyte tapes are only 64 bytes in length, despite what the standard says about block size being a minimum of 128 bytes.

5 *Main stream.* Contains the main data to be stored on a part of the CD, treated simply as a stream of bytes, irrespective of the block or packet size used. More than one of these files can be used in cases of mixed-mode disks, but there is normally only one in the case of a conventional audio CD. If stored as a file, its name is indicated in the appropriate map packet.

DDP tapes for audio CDs written in the order shown above make it impossible to modify PQ (subcode) data once the tape has been finished, unless the whole thing is rewritten, including the audio data. This is because the subcode stream is sandwiched between the map stream and the audio. For this reason, some workstation software will allow the streams to be written in a different order, as shown in Figure 6.6, whereby the audio comes first and the subcode last. Although non-standard,

Figure 6.6 Order of streams on a DDP master tape for audio CD. (a) Conventional order. (b) Modified order allowing subcode updates

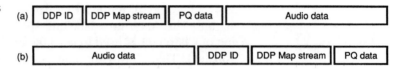

| (a) | DDP ID | DDP Map stream | PQ data | Audio data |

| (b) | Audio data | DDP ID | DDP Map stream | PQ data |

many CD cutters will apparently accept the data in this form, and it gives greater flexibility to the mastering engineer.

Because the transfer rate of Exabyte tapes is quite high, CD glass masters can be cut from Exabyte/DDP tapes at rates considerably faster than real time. At the time of writing it is possible to cut a CD at approximately 2.5 times real time, which is one very persuasive reason for using it in preference to alternative systems. As the use of networks increases, it is possible that users may begin to transfer CD masters using DDP without the need for physical media.

6.4 Digital audio interfaces

It is often necessary to interconnect audio workstations so that digital audio data can be transferred between them without converting back to analog form. This preserves sound quality, and is normally achieved using one of the standard point-to-point digital interfaces described below. These are different from computer data networks in that they are designed purely for the purpose of carrying audio data in real time, and cannot be used for general purpose file transfer applications.

6.4.1 Computer networks and digital audio interfaces compared

Real-time audio interfaces are the digital audio equivalent of signal cables, down which digital audio signals for one or more channels are carried in real time from one point to another, possibly with some auxiliary information attached. A real-time audio interface uses a data format dedicated specifically to audio purposes, unlike a computer data network which is not really concerned with what is carried in its data packets. A recording transferred over a digital interface to a second machine may be copied 'perfectly' or cloned, and this process takes place in real time, requiring the operator to put the receiving device into record mode such that it simply stores the incoming stream of audio data. The auxiliary information may or may not be recorded (usually most of it is not).

Real-time interfaces are normally unidirectional, point-to-point connections, and should be distinguished from buses and computer networks which are often bidirectional and which carry data in a packet format. Sources may be connected to destinations using a routing matrix, very much as with analog signals. Audio data is transmitted in an unbroken stream, there is no handshaking process involved in the data transfer, and erroneous data may not be retransmitted – there is no mechanism for requesting its retransmission. The data rate of a real-time audio interface is directly related to the sampling rate, wordlength and number of channels of the audio data to be transmitted, thus ensuring that the interface is always capable of serving the specified number of channels. If a channel is unused for some reason its capacity is not normally available for assigning to other purposes (such as higher speed transfer of another channel, for example).

Clearly real-time interfaces are best suited to operational situations in which analog signal cabling needs to be replaced by a digital equivalent, and where digital audio signals are to be routed from place to place within a studio environment so as to ensure dedicated signal feeds.

In contrast, a computer network typically operates asynchronously and data is transferred in the form of packets. One would expect a number of devices to be interconnected using a single network, and for an addressing structure to be used such that data might be transferred from a certain source to a certain destination. Bus arbitration is used to determine the existence of network traffic, and to avoid conflicts in bus usage. The 'direction' of data flow is determined simply by the source and destination addresses. The speed of the network effectively determines the amount of traffic which may be present on the bus and the rate at which it will be transferred, this bearing little or no relationship to the audio sample rate, the number of channels or the wordlength. The format of data on the network is not necessarily audio specific (although protocols optimised for audio transfer may be used), and one network may carry text data, graphics data and E-mail, all in separate packets, for example.

6.4.2 Interconnection in the analog domain

In the case of analog interconnection, replayed audio is converted back to the analog domain by the replay machine's D/A convertors, routed to the recording machine via a conventional audio cable, and then reconverted to the digital domain by the recording machine's A/D convertors. The audio is subject to any gain changes which might be introduced by level differences between output and input, or by the record gain control of the recorder and the replay gain control of the player. Analog domain copying is necessary if any analog processing of the signal is to happen in between player and recorder, such as gain correction, equalisation, or the addition of effects such as reverberation. More and more of these operations, though, are now possible in the digital domain.

An analog domain copy cannot be said to be a perfect copy or a clone of the original master, since the data values will not be exactly the same (due to slight differences in recording level, differences between convertors, etc.) – for a clone it is necessary to make a true digital copy.

6.4.3 Interconnection in the digital domain

Professional digital audio systems, and some consumer systems, have digital interfaces conforming to one of the standard protocols and allow for a number of channels of digital audio data to be transferred between devices with no loss of sound quality. Any number of generations of digital copies may be made without affecting the sound quality of the latest generation, provided that errors have been fully corrected. The digital outputs of a recording device are taken from a point in the signal chain after error correction, which results in the copy being error corrected. Thus the copy does not suffer from any errors which existed in the master, provided that those errors were correctable.

Digital interfaces may be used for the interconnection of recording systems and other digital audio devices such as mixers and effects units. It will eventually become common only to use analog interfaces at the very beginning and end of the signal chain, with all other interconnections being made digitally. A number of installations now exist which use all digital interconnection for audio.

6.4.4 Digital interface types

There are a number of types of digital interface, some of which are international standards and others which are manufacturer-specific. They all carry digital audio with at least 16 bit resolution, and will operate at the standard sampling rates of 44.1 and 48 kHz, as well as at 32 kHz if necessary, with a degree of latitude for varispeed. Most of the interface standards are one- or two-channel only, but one of them, known commonly as MADI, is a multichannel interface.

Manufacturer-specific interfaces may also be found on equipment from other manufacturers if they have been considered necessary for the purposes of signal interchange, especially on devices manufactured prior to the standardisation and proliferation of the AES/EBU interface. The interfaces vary as to how many physical interconnections are required, since some require one link per channel plus a synchronisation signal, while others carry all the audio information plus synchronisation information over one cable.

Making a copy of a recording using any of the interface standards involves the connection of the appropriate cables between player and recorder, and the switching of the recorder's input to 'digital' as opposed to 'analog', since this sets it to accept a signal from the digital input as opposed to the A/D convertor. It is necessary for both machines to be operating at the same sampling rate (unless a sampling rate convertor is used) and may require the recorder to be switched to 'external sync' mode, so that it can lock its sampling rate to that of the player. Alternatively (and preferably) a common reference signal may be used to synchronise all devices which are to be interconnected digitally. A recorder should be capable of at least the same quantising resolution (number of bits per sample) as a player, otherwise audio resolution will be lost. If there is a difference in resolution between the systems it is advisable to use a processor in between the machines which optimally dithers the signal for the new resolution, or alternatively to use redithering options on the source machine to prepare the signal for its new resolution.

The interfaces are described below in outline. It is common for subtle incompatibilities to arise between devices, even when interconnected with a standard interface, owing to the different ways in which non-audio information is implemented. This can result in anything from minor operational problems to total non-communication, and the causes and remedies are unfortunately far too detailed to go into here. The reader is referred to *The Digital Interface Handbook* by Rumsey and Watkinson,[8] as well as to the standards themselves, if a greater understanding of the intricacies of digital audio interfaces is required.

6.4.5 The AES/EBU interface (AES3-1992)

The AES/EBU interface,[9] described almost identically (but not quite) in AES3-1992, IEC 958 (Type 1), CCIR Rec 647 and EBU

Figure 6.7 Recommended electrical circuit for use with the standard two-channel interface

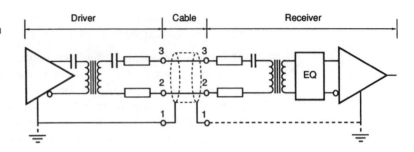

Tech. 3250E, allows for two channels of digital audio (A and B) to be transferred serially over one balanced interface, using balanced drivers and receivers similar to those used in the RS422 data transmission standard, with an output voltage of between 2 and 7 volts, as shown in Figure 6.7. The interface allows two channels of audio to be transferred over distances up to 100 m, but longer distances may be covered using combinations of appropriate cabling, equalisation and termination. Standard XLR-3 connectors are used, often labelled DI (for digital in) and DO (for digital out).

Each audio sample is contained within a 'subframe' (see Figure 6.8), and each subframe begins with one of three synchronising patterns to identify the sample as either A or B channel, or to mark the start of a new channel status block (see Figure 6.9). These synchronising patterns violate the rules of bi-phase mark coding (see below), and thus are easily identified by a decoder. Additional data is carried within the subframe in the form of 4 bits of auxiliary data (which may either be used for additional audio resolution or for other purposes such as low-quality speech), a validity bit (V), a user bit (U), a channel status bit (C) and a parity bit (P), making 32 bits per subframe and 64 bits per frame.

One frame (containing two audio samples) is transmitted in the time period of one audio sample, and thus the data rate varies

Figure 6.8 Format of the standard two-channel interface frame

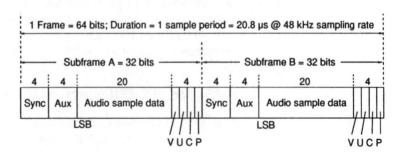

Figure 6.9 Three different preambles (X, Y and Z) are used to synchronise a receiver at the starts of subframes

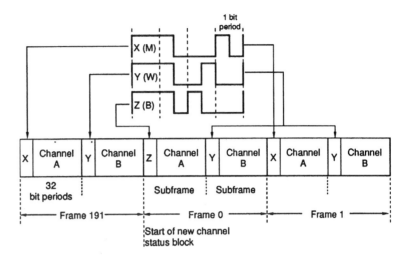

Figure 6.10 Channel status bits, corrected over 192 frames, are formed into a 24 byte word. The functions of each byte are shown here.

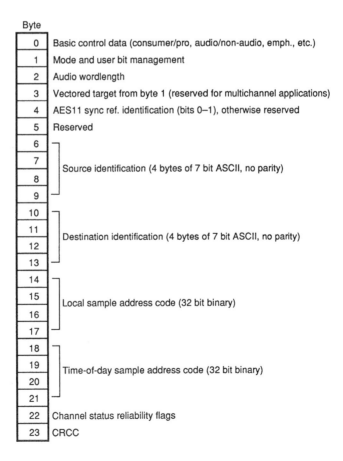

Byte	
0	Basic control data (consumer/pro, audio/non-audio, emph., etc.)
1	Mode and user bit management
2	Audio wordlength
3	Vectored target from byte 1 (reserved for multichannel applications)
4	AES11 sync ref. identification (bits 0–1), otherwise reserved
5	Reserved
6	
7	Source identification (4 bytes of 7 bit ASCII, no parity)
8	
9	
10	
11	Destination identification (4 bytes of 7 bit ASCII, no parity)
12	
13	
14	
15	Local sample address code (32 bit binary)
16	
17	
18	
19	Time-of-day sample address code (32 bit binary)
20	
21	
22	Channel status reliability flags
23	CRCC

Figure 6.11 An example of the biphase-mark channel code

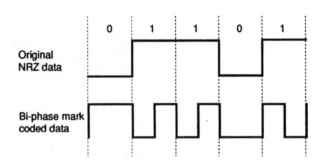

with the sampling rate. Channel status bits are aggregated at the receiver to form a 24 byte word every 192 frames, and each bit of this word has a specific function relating to interface operation, an overview of which is shown in Figure 6.10. Examples of bit usage in this word are the signalling of sampling rate and pre-emphasis, as well as the carrying of a sample address 'timecode' and labelling of source and destination. Bit 1 of the first byte signifies whether the interface is operating according to the professional (set to 1) or consumer (set to 0) specification.

Bi-phase mark coding, the same channel code as used for SMPTE/EBU timecode, is used in order to ensure that the data is self-clocking, of limited bandwidth, DC free, and polarity independent, as shown in Figure 6.11. The interface has to accommodate a wide range of cable types, and a nominal 110 ohms characteristic impedance is recommended. Originally (AES3-1985) up to four receivers with a nominal input impedance of 250 ohms could be connected across a single professional interface cable, but a more recent modification to the standard recommended the use of a single receiver per transmitter, having a nominal input impedance of 110 ohms.

Figure 6.12 The consumer electrical interface

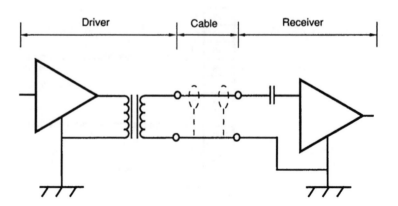

6.4.6 Standard consumer interface (IEC 958, Type 2)

The consumer interface (historically related to SPDIF – the Sony/Philips digital interface) is very similar to the professional AES/EBU interface, but uses unbalanced electrical interconnection over a coaxial cable having a characteristic impedance of 75 ohms, as shown in Figure 6.12. It can be found on many items of semi-professional or consumer digital audio equipment, such as CD players and DAT machines. It usually terminates in an RCA phono connector, although some hi-fi equipment makes use of optical fibre interconnects carrying the same data. Format convertors are available for converting consumer format signals to the professional format, and vice versa, and for converting between electrical and optical formats.

When the IEC standardised the two-channel digital audio interface, two requirements existed: one for 'consumer use', and one for 'broadcasting or similar purposes'. A single IEC standard (IEC 958) has resulted with only subtle differences between

Figure 6.13 Functions of channel status bytes in consumer interface

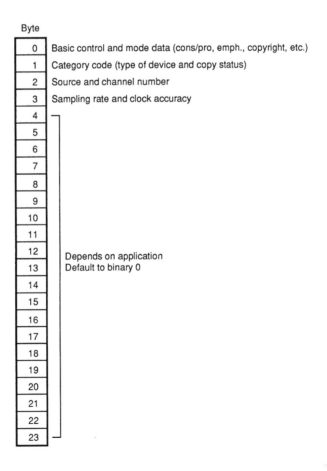

Byte	
0	Basic control and mode data (cons/pro, emph., copyright, etc.)
1	Category code (type of device and copy status)
2	Source and channel number
3	Sampling rate and clock accuracy
4	
5	
6	
7	
8	
9	
10	
11	
12	Depends on application
13	Default to binary 0
14	
15	
16	
17	
18	
19	
20	
21	
22	
23	

consumer and professional implementation. Occasionally this causes problems in the interconnection of machines, such as when consumer format data is transmitted over professional electrical interfaces. IEC 958 is currently under revision.

The data format of subframes is the same as that used in the professional interface, but the channel status implementation is almost completely different, as shown in Figure 6.13. The second byte of channel status in the consumer interface has been set aside for the indication of 'category codes', these being set to define the type of consumer usage. Current examples of defined categories are (00000000) for the General category, (10000000) for Compact Disc and (11000000) for a DAT machine. Once the category has been defined, the receiver is expected to interpret certain bits of the channel status word in different ways, depending on the category. For example, in CD usage, the four control bits from the CD's 'Q' channel subcode are inserted into the first four control bits of the channel status block (bits 1–4). Copy protection can be implemented in consumer-interfaced equipment, according to the Serial Copy Management System (SCMS).[10]

The user bits of the consumer interface are often used to carry information derived from the subcode of recordings, such as track identification and cue point data. This can be used when copying CDs and DAT tapes, for example, to ensure that track start ID markers are copied along with the audio data. This information is not normally carried over AES/EBU interfaces.

6.4.7 Manufacturers' interfaces

The most common manufacturer-specific interface has been Sony's SDIF-2, which was designed for the transfer of one channel of digital audio information per cable, at a resolution of up to 20 bits (although most devices only make use of 16). The interface, as used on most two-channel equipment, is unbalanced and uses 75 ohm coaxial cable terminating in 75 ohm BNC-type connectors, one for each audio channel. TTL-compatible electrical levels (0–5 V) are used. The audio channel connectors are accompanied by a word clock signal on a separate connector, which is a square wave at the sampling frequency used to synchronise the receiver's sample clock. There is also a multichannel electrical interface conforming to RS422 standards and using D-type multiway connectors. A single BNC connector carries word clock as before.

In each audio sample period, the equivalent of 32 bits of data is transmitted over each interface, although only the first 29 bits of the word are considered valid, since the last three bit-cell periods are divided into two cells of one-and-a-half times the normal duration, in order to act as a synchronising pattern. Audio data are transmitted with the MSB first, followed by nine control or user bits. The resulting data rate is 1.53 Mbit/s at 48 kHz sampling rate and 1.21 Mbit/s at 44.1 kHz.

The SDIF-2 interface was used mainly for the transfer of audio data from Sony professional digital audio equipment, particularly the PCM-1610 and 1630 CD-mastering PCM adaptors. It may also appear on some other items of professional audio equipment for the purposes of interfacing to Sony equipment.

A number of other manufacturer interfaces have arisen from time to time, particularly on low-cost digital audio equipment, including examples from Yamaha and Tascam. Commercial interface convertors are available which simplify the interconnection of such devices with others using standard interfaces. It is not proposed to cover these in any more detail here.

6.4.8 Standard multichannel interface (AES10–1991)

A number of digital audio manufacturers combined forces to propose a multi-channel serial interface standard (AES10-1991[11]) known as MADI (the Multi-channel Audio Digital Interface), which is based on the two-channel AES/EBU interface. It has been designed to be transparent to AES/EBU data and has applications in large-scale digital routing systems and interconnection of multichannel audio equipment. MADI naturally uses a higher data rate to carry the increased amount of information and allows for 56 channels of audio to be carried serially over one 75 ohm coaxial cable or optical fibre, such that one sample for each channel is transferred within one audio sample period.

Allowing for maximum commonality with the two-channel AES/EBU data format, MADI allows for subframes containing 20 or 24 bit audio data, together with the additional status bits of the AES/EBU data for each channel. The main difference between MADI and AES/EBU is in the first 4 bits of each subframe (which in the two-channel interface break the bi-phase mark coding law to provide a sync pattern, but are used for other header information in MADI). Channel subframes are linked sequentially to form a 56 channel frame, as shown in Figure 6.14.

The transmission data rate of MADI is fixed at 125 Mbit/s, irrespective of the sampling rate or number of channels, and the actual data transfer rate is 100 Mbit/s due to the use of a 4–5 bit

Figure 6.14 Format of the
MADI frame

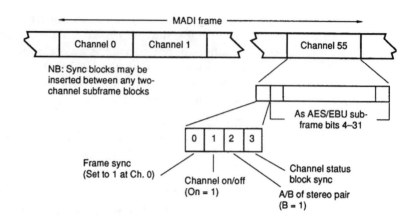

encoding scheme. In this channel code each 32 bit subframe is
broken up into 4 bit words which are then encoded to 5 bit
words according to a look-up table, the reason being to maintain
a low DC content to the code. A synchronisation symbol (11000
10001) is inserted at least once per frame, and in cases where the
full bandwidth of the link is not being used, additional sync
symbols are inserted to take up the capacity of the bus. A typical
MADI link configuration is shown in Figure 6.15.

Communication is expected to be handled entirely by so-called
TAXI (Transparent Asynchronous Xmitter/Receiver Interface)
chips, which automatically take care of the insertion of sync
symbols. Unlike the two-channel format, it is intended that the
link will be asynchronous and that transmitter and receiver will
be locked to a common synchronising clock (in the form of an
AES/EBU reference signal). BNC 75 ohm connectors are to be
used, and the maximum coaxial cable length is intended to be no
more than 50 metres (although optical fibre would allow more,
this being an alternative means of interconnection). The modula-

Figure 6.15 Block diagram
of MADI transmission and
reception

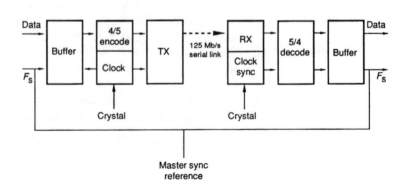

tion method used is NRZI in which a transition from high-to-low or low-to-high represents a binary 'one', and no transition represents a zero. Because of the asynchronous nature of the interface, buffers are required at both ends of the link so that data can be re-clocked from the buffer at the correct rate. At the receiver, data is clocked out under control of the synchronising signal.

6.4.9 Operational problems

If a digital interface between two devices appears not to be working it could be due to one or more of the following conditions. A number of bench and hand-held digital audio interface

Figure 6.16 Example of a hand held interface analyser, which displays the states of certain bits commonly causing device incompatibility and allows them to be modified. It also has outputs to a scope for waveform display and a small monitor loudspeaker. (Courtesy of Audio Digital Technology)

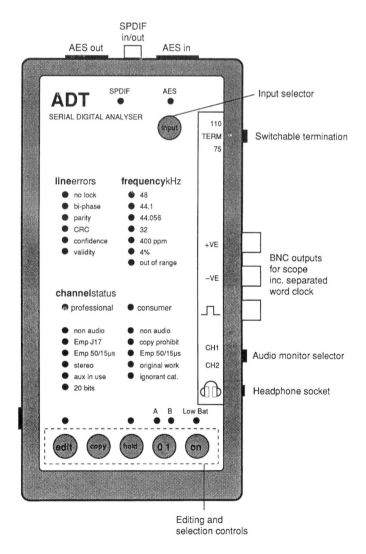

analysers are available, such as the one pictured in Figure 6.16, which can help in working out which aspects of the interface or the data format are at fault.

Asynchronous sample rates

The two devices must normally operate at the same sample rate, preferably locked to a common reference. Ensure that the receiver is in external sync mode and that a synchronizing signal (common to the transmitter) is present at the receiver's sync input. Alternatively, set the receiver to lock to the digital audio input (standard two-channel interfaces only). If the incoming signal is at the wrong sampling rate it must be resynchronized or sample rate converted.

'Sync' or 'locked' indicator flashing or out on the receiver normally means that no external sync reference exists or that it is different from that of the signal at the digital input. Check that external sync reference and digital input are at the correct rate and locked to the same source. Decide on whether to use internal or external sync reference, depending on application.

If problems with 'good lock' or drifting offset arise when locking to other machines or when editing, check that any timecode is synchronous with video frame rate and audio sampling rate.

Digital input

It may be that the receiver is not switched to accept a digital input.

Data format

Received data is in the wrong format. Both transmitter and receiver must operate to the same format. Conflicts may exist in such areas as channel status, and there may be a consumer–professional conflict. Use a format convertor to set the necessary flags, or alter settings in the transmitter or receiver.

Cables and connectors

Cables or connectors may be damaged or incorrectly wired. Cable may be too long, of the wrong impedance, or generally of poor quality. Digital signal may be of poor quality. Interface analysers are available which give an indication of the electrical quality of the signal.

SCMS (consumer interface only)

Copyright or SCMS flag may be set by the transmitter. For professional purposes, use a format convertor to set the necessary flags or use the AES/EBU interface which is not subject to SCMS.

Receiver mode

Receiver is not in record or input monitor mode. Some recorders must be at least in record–pause before they will give an audible and metered output derived from a digital input.

6.5 Digital signal synchronisation

6.5.1 *The need for synchronisation*

Unlike analog audio, digital audio has a discrete-time structure, because it is a sampled signal in which the samples may be further grouped into frames and blocks having a certain time duration. If digital audio devices are to communicate with each other, or if digital signals are to be combined in any way, then it is important that they are synchronized to a common reference in order that the sampling frequencies of the devices are identical and do not drift with relation to each other. It is not enough for two devices to be running at nominally the same sampling frequency (say, both at 44.1 kHz), since between the sampling clocks of professional audio equipment it is possible for differences in frequency of up to ±10 parts per million (ppm) to exist, and even a very slow drift means that two devices are not truly synchronous.

The audible effect resulting from a non-synchronous signal drifting with relation to a sync reference or another signal is usually the occurrence of a glitch or click at the difference frequency between the signal and the reference, typically at an audio level around 50 dB below the signal, due to the repetition or dropping of samples. This will appear when attempting to mix two digital audio signals whose sampling rates differ by a small amount, or when attempting to decode a signal such as an unlocked consumer source by a professional system which is locked to a fixed reference. This said, it is not always easy to detect asynchronous operation by listening, even though sample slippage is occurring. Some systems may not operate at all if presented with asynchronous signals.

Furthermore, when digital audio is used with analog or digital video, the sampling rate of the audio needs to be locked to the video reference signal, and also to any timecode signals which may be used. In single studio operations the problem of ensuring lock to a common clock is not as great as it is in a multi-studio centre, or where digital audio signals arrive from remote locations. In such cases either the remote signals must be synchronized to the local sample clock as they arrive, or the remote studio must somehow be fed with the same reference signal as the local studio.

6.5.2 Choice of sync reference

There are now AES recommendations for the synchronization of digital audio signals, documented in AES11-1991.[12] They state that preferably all machines should be able to lock to a reference signal, which should take the form of a standard two-channel interface signal (see section 6.3) whose sampling frequency is stable within a certain tolerance, and that all machines should have a separate input for such a synchronizing signal. If this procedure is not adopted then it is possible for a device to lock to the clock embedded in the channel code of the AES-format audio input signal – a technique known as 'genlock' synchronization.

In the AES11 standard signals are considered synchronous when they have identical sample rates, but phase errors are allowed to exist between the reference clock and received/transmitted digital signals in order to allow for effects such as cable propagation delays, phase-locked loop errors and other electrical effects. Input signal frame edges must lie within ±25 per cent of the reference signal's frame edge (taken as the leading edge of the 'X' preamble), and output signals within ±5 per cent (see Figure 6.17), although tighter accuracy than this is preferable because otherwise an unacceptable build up of delay may arise when equipment is cascaded.

Figure 6.17 Timing limits recommended for synchronous signals in the AES11 standard

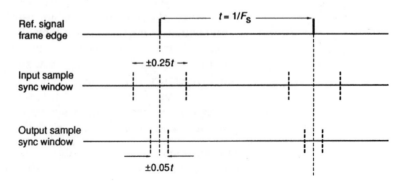

The AES11 reference signal may either contain program or not. If it does not contain program it may be digital silence or simply the sync preamble with the rest of the frame inactive. Two grades of reference are specified: Grade 1, having a long-term frequency accuracy of ±1 ppm, and Grade 2, having a long-term accuracy of ±10 ppm. The Grade 2 signal is intended for use within a single studio which has no immediate technical reason for greater accuracy, whereas Grade 1 is a tighter specification and is intended for the synchronization of complete studio centres (as well as single studios if required). These AES recommendations, of which this is only a summary, were only defined in 1991 and equipment manufacturers have so far been slow to adopt the provision of a separate AES-format input for a synchronizing signal. It is expected that this will become more common over the next few years.

Digital audio workstations designed for professional use are often provided with a wide range of sync inputs, and most systems may be operated in the external or internal sync modes. In the internal sync mode a system is locked to its own crystal oscillator, which in professional equipment should be accurate within ±10 ppm if it conforms to AES recommended practice (AES5-1984), but which in consumer equipment may be much less accurate than this.

In the external sync mode the system should lock to one of its sync inputs, which may either be selectable using a switch, or be selected automatically based on an order of priority depending on the mode of operation (for this the user should refer to the operations manual of the device concerned). Typical sync inputs are word clock (WCLK), which is normally a square-wave TTL-level signal (0–5 V) at the sampling rate, usually available on a BNC-type connector and common on equipment using the Sony interface (SDIF), and 'composite video', which is a video reference signal consisting of either normal picture information or just 'black and burst' (a video signal with a blacked-out picture). In all cases one machine or source must be considered to be the 'master', supplying the sync reference to the whole system, and the others as 'slaves'.

WCLK may be 'daisy-chained' (looped through) between devices in cases where the AES/EBU interface is not available. A 'sync' light is usually provided on the front panel of a device (or under a cover) to indicate good lock to an external or internal clock, and this may flash or go out if the system cannot lock to the clock concerned, perhaps because it has too much jitter, is at too low a level, conflicts with another clock, or because it is not at a sampling rate which can be accepted by the system.

Video sync (composite sync) is often used when a system is operated within a video environment, and where a digital audio system is to be referenced to the same sync reference as the video machines in a system.

6.5.3 Distribution of sync references

It is appropriate to consider digital audio as similar to video when approaching the subject of sync distribution, especially in large systems. Consequently, it is advisable to use a central high quality sync signal generator (the equivalent of a video sync pulse generator, or SPG), the output of which is made available widely around the studio using distribution amplifiers to supply different outlets in a 'star' configuration. In video operations this is often called 'house sync', and such a term may become used in digital audio as well. Each digital device in the system may then be connected to this house sync signal, and each should be set to operate in the external sync mode. Until such time as AES11 reference inputs become available on audio products one might use word clock or a video reference signal instead.

Using the technique of central sync-signal distribution (see Figure 6.18) it becomes possible to treat all devices as slaves to the sync generator, rather than each one locking to the audio output of the previous one. In this case there is no 'master' machine, since the sync generator acts as the 'master'. It requires that all machines in the system operate at the same sampling rate, unless a sample rate convertor or signal synchronizer is used.

Figure 6.18 The best synchronisation strategy usually involves distribution of one central sync reference to a separate sync input on all devices, either in the form of a video sync reference, an AES 11 reference, a word clock reference or other proprietary sync signal

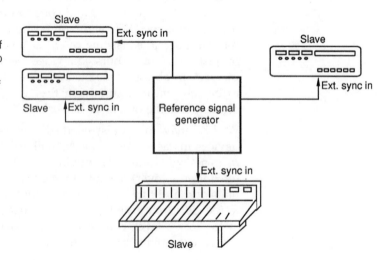

Figure 6.19 Alternative digital sync strategies. (a) One studio device with a stable clock is used as a reference by all the other devices. (b) Devices lock to their AES/EBU audio inputs (often only usable when recording on these devices, since sync replay normally requires the use of a separate sync input, unless the device has been modified to lock to its AES input during replay)

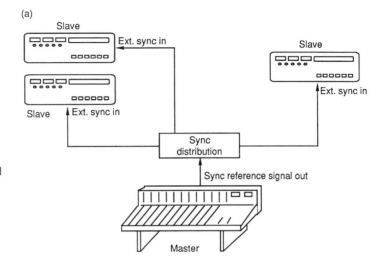

(a)

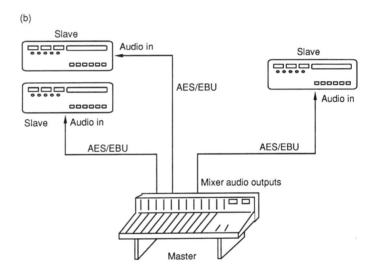

(b)

Alternatively, in a small studio, it may be uneconomical or impractical to use a separate SPG, and in such cases one device in the studio must be designated as the master. This device would then effectively act as the SPG, operating in the internal sync mode, with all other devices operating in the external sync mode and slaving to it (see Figure 6.19). If a digital mixer were to be used then it could be used as the SPG, but alternatively it would be possible to use a tape recorder, workstation or other device with a stable clock.

6.5.4 External sync and sound quality

As discussed in section 2.2, the stable timing of digital audio samples is crucial to the quality of conversion back to the analog domain. Since the clock used in conversion must run at the same sampling rate as the rest of the workstation it is normal to derive this clock from whatever synchronisation reference is in use at the time. Internal clock signals used in workstations can be made quite accurate, and normally have good stability, but external sync signals can be less stable, depending on factors such as the length of cable over which they have travelled, external interference and stability of the sync generator. It is common for cheaper systems to pass on external clock instabilities to the convertors, leading to poorer sound quality, whereas the correct approach is to filter the external clock signal to remove as much jitter as possible.

6.5.5 Considerations in video environments

In environments where digital audio is used with video signals, it is important for the audio sampling rate to be locked to the same master clock as the video reference. The same applies to timecode signals which may be used with digital audio and video equipment.

People using the PAL or SECAM television systems (see Chapter 7) are fortunate in that there is a simple integer relationship between the sampling rate of 48 kHz used in digital audio systems for TV and the video frame rate of 25 Hz (there are 1920 samples per frame). There is also a simple relationship between the other standard sampling rates of 44.1 and 32 kHz and the PAL/SECAM frame rate, as shown in Table 6.1. Users of NTSC TV systems (such as the USA and Japan) are less fortunate because the TV frame rate is 30/1.001 (roughly 29.97) frames per second, resulting in a non-integer relationship with standard audio sampling rates. The sampling rate of 44.056 kHz was introduced in digital audio recording systems which used NTSC VTRs, since this resulted in an integer relationship with the frame rate, but this sampling rate is used infrequently today.

Table 6.1 Audio samples per TV frame in PAL and NTSC systems

Sampling rate	PAL/SECAM	NTSC
32 kHz	1280	16016/15
44.1 kHz	1764	147147/100
48 kHz	1920	8008/5

Figure 6.20 In video
environments all audio, video
and timecode sync
references should be locked
to a common clock

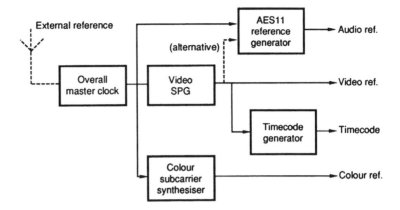

It is desirable to source a master audio reference signal centrally
within a studio operation, just as a video reference is centrally
sourced, and these two references are normally locked to the
same highly stable rubidium reference, which in turn may be
locked to a standard reference frequency broadcast by a national
transmitter. As shown in Figure 6.20, the overall master clock
should be used to lock the video SPG, any audio reference signal
generators and the video colour subcarrier synthesizer. The
video reference should also be used to lock timecode generators.
Appropriate master clock frequencies and division ratios may be
devised for the application in question.

6.6 Networks

General purpose data networks are increasingly used for digital
audio, not as signal cables but as a means of transferring files
between distributed equipment. Not only can networks be used
to link a number of local workstations and storage resources, but
they can also be used to transfer files over very large distances,
even across the world if necessary. This has made it possible for
people to work on common material, using shared resources
such as storage and backup, and sometimes to avoid the time
delay and expense of physically transporting recording media to
other sites.

As with storage devices, it is only recently that networks have
become fast enough and cheap enough to be viable in this application, since, as already explained, digital audio requires considerable bandwidth for real time transfer. If networks are to be
truly useful for digital audio applications then they really need
to be capable of handling the file transfer for a number of
channels of audio simultaneously, and preferably at speeds in

excess of real time. It may, nevertheless, be acceptable in some circumstances for a network to transfer in non-real time, especially if one is dealing with an international transfer of audio that would have taken longer or cost more if one had had to physically transport the recording instead.

6.6.1 Network principles

A network carries data either on wire or optical fibre, and is normally shared between a number of devices and users. The sharing is achieved by containing the data in packets of a limited number of bytes (usually between 64 and 1518), each with an address attached, rather like MIDI data (see Chapter 8). The packets may share a common physical link, which is normally a high speed serial bus of some kind, being multiplexed in time either using a regular time slot structure or in an asynchronous fashion whereby the time interval between packets may be varied, as shown in Figure 6.21. The length of packets also may not be constant, depending on the requirements of different protocols sharing the same network. Packets for a particular file transfer between two devices may not be contiguous and may be transferred erratically, depending on what other traffic is sharing the same physical link.

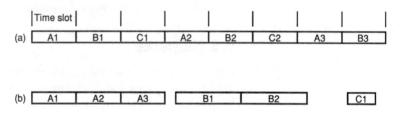

Figure 6.21 Packets for different destinations (A, B and C) multiplexed onto a common serial bus. (a) Time division multiplexed into a regular time slot structure. (b) Asynchronous transfer showing variable time gaps and packet lengths between transfers for different destinations

Figure 6.22 shows some common physical layouts for local area networks (LANs). LANs are networks which operate within a limited area, such as an office building or studio centre, within which it is common for every device to 'see' the same data, each picking off that which is addressed to it and ignoring the rest. In order to place a packet of data on the network, devices must have a means for determining whether the network is busy, and there are various protocols in existence for arbitrating network access. In the 'backbone' configuration devices are connected to spurs off a common serial bus which requires the bus to be 'chained' between each successive device. Here, a break in the chain can mean disconnection for more than one device. The star configuration involves a central hub which distributes the data

Figure 6.22 Two examples of computer network topologies. (a) Devices connected by spurs to a common hub, and (b) devices connected to a common 'backbone'

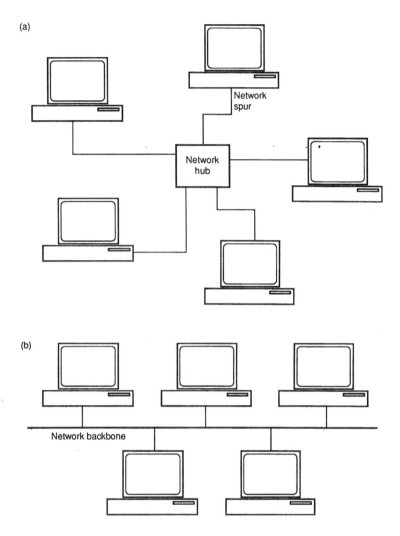

to each device separately, and this is more reliable because a break in any one link does not affect the others. Bus arbitration in both these cases is normally performed by collision detection which is a relatively crude approach, relying very much on the rules of spoken conversation between people. Devices attempt to place packets on the bus whenever it appears to be quiet, but a collision may take place if another device attempts to transmit before the first one has finished. The network interface of the transmitting device detects the collision by attempting to read the data it has just transmitted and retransmits it after transmitting a brief 'blocking signal' if it has been corrupted by the collision.

A token ring configuration places each device within a 'ring', each having both an 'in' and an 'out' port, with bus arbitration performed using a process of token passing from one device to the next. This works rather like trains running on a single track line, in that a single token is carried by the train using the line, and trains can only use the line if carrying the token. The token is passed to the next train upon leaving the single track sector to show that the line is clear.

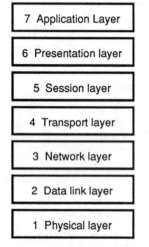

| 7 Application Layer |
| 6 Presentation layer |
| 5 Session layer |
| 4 Transport layer |
| 3 Network layer |
| 2 Data link layer |
| 1 Physical layer |

Figure 6.23 The ISO model for Open Systems Interconnection is arranged in seven layers, as shown here

Network communication is divided into a number of 'layers', each of which determines an aspect of communication. The ISO seven-layer model for open systems interconnection (OSI) shows how many levels there are at which compatibility between systems needs to exist before seamless interchange of data can be achieved (Figure 6.23). It shows that communication begins with the application and filters down through various stages to the layer most people understand – the physical layer, or the piece of wire over which the information is carried. Layers 3, 4 and 5 can be grouped under the broad heading of 'protocol', determining the way in which data packets are formatted and transferred. There is a strong similarity here with the exchange of data on physical media, as discussed earlier, where a range of compatibility layers from the physical to the application determine whether or not one device can read another's disks.

6.6.2 Network standards

Ethernet, FDDI (Fibre Distributed Data Interface) and ATM (Asynchronous Transfer Mode) are examples of network standards, each of which specifies a number of layers within the OSI model. FDDI, for example, specifies only the first three layers of the OSI model (the physical, data link and network layers).

Ethernet allows a number of different methods of interconnection and runs at a maximum rate of 10 Mbit/s (although there is now a version running at 100 Mbit/s), using collision detection for network access control. Interconnection can be via either thick (10-Base-10) or thin (10-Base-2) coaxial cable, the thin being quite common and normally working in the backbone-type configuration shown in the previous section, using 50 ohm BNC connectors and T-pieces to chain devices on the network (see Figure 6.24). Such a configuration requires resistive terminators at the ends of the bus to avoid reflections, as with SCSI connections. Twisted-pair (10-Base-T) connection is also becoming quite popular, and is often configured in the star topology using a central hub. Devices can then be plugged and unplugged without affecting others on the network.

Figure 6.24 Typical thin Ethernet interconnection arrangement

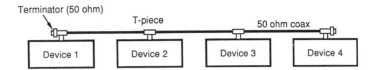

FDDI[13] is a high speed optical fibre network running at 100 Mbit/s, and operating on the token passing principle described above, allowing up to 2 km between stations. There is also a copper version of FDDI called CDDI which runs at the same rate but restricts interconnection distance. ATM is really only a protocol for data communication, which does not specify the physical medium for interconnection. Physical systems already exist which are capable of running at 250 Mbit/s, and plans are in hand for systems running at 2.4 Gbit/s. Both FDDI and ATM have been considered quite suitable for audio and other multi-media purposes, and special protocols have been devised by some manufacturers to optimise the transfer bandwidth for real-time recording and replay purposes.

6.6.3 Network protocols

A protocol specifies the format of communication on a network. In other words it determines things like the format of data packets, their header information and addressing structure, and any handshaking and error retrieval schemes, amongst other things. One physical network can handle a wide variety of protocols, and packets conforming to different protocols can coexist on the same bus.

Some common examples of general purpose network protocols are such as AppleTalk, used for file sharing and printing between Macintosh computers; TCP/IP, used for communications over the Internet (see below); and Novell, used widely in PC networks. These general purpose protocols are not particularly efficient for real-time audio transfer, but they can be used for non-real-time transfer of audio files between workstations. Specially designed protocols may be needed for audio purposes, as described below.

6.6.4 Audio network requirements

The principal application of computer networks in audio systems is in the transfer of audio data files between workstations, or between workstations and a central 'server' which stores shared files. The device requesting the transfer is known as the 'client' and the device providing the data is known as the

209

'server'. When a file is transferred in this way a byte-for-byte copy is reconstructed on the client machine, with the file name and any other header data intact. There are considerable advantages in being able to perform this operation at speeds in excess of real time for operations in which real-time feeds of audio are not the aim. For example, in a news editing environment a user might wish to load up a news story file from a remote disk drive in order to incorporate it into a report, this being needed as fast as the system is capable of transferring it. Alternatively, the editor might need access to remotely stored files, such as sound files on another person's system, in order to work on them separately. In audio post-production for films or video there might be a central store of sound effects, accessible by everyone on the network, or it might be desired to pass on a completed portion of a project to the next stage in the post-production process.

Simple desktop computer networks such as Apple's LocalTalk are very slow by audio standards and are not capable of transferring even a single channel sound file at real-time speeds, whereas Ethernet is fast enough to transfer audio data files slightly faster than real time, depending on network loading. Approaches using FDDI or ATM over optical fibre are most appropriate for handling large numbers of sound file transfers simultaneously at high speed. Unlike a real-time audio interface, the speed of transfer of a sound file over a network (when using conventional file transfer protocols) depends on how much traffic is currently using it. If there is a lot of traffic then the file may be transferred more slowly than if the network is quiet (very much like motor traffic on roads). The file might be transferred erratically as traffic volume varies, with the file arriving at its destination in 'spurts'. There therefore arises the need for network communication protocols designed specifically for the transfer of real-time data, which serve the function of reserving a proportion of the network bandwidth for a given period of time, as described below.

Without real-time protocols designed as indicated above, the computer network may not be relied upon for transferring audio where an unbroken audio output is to be reconstructed at the destination from the data concerned. The faster the network the more likely it is that one would be able to transfer a file fast enough to feed an unbroken audio output, but this should not be taken for granted. Even the highest speed networks can be filled up with traffic! This may seem unnecessarily careful until one considers an application in which a disk drive elsewhere on the network is being used as the source for replay by a local

Figure 6.25 In this example of a networked system a remote disk is accessed over the network to provide data for real-time audio playout from a workstation used for on-air broadcasting. Continuity of data flow to the on-air workstation is of paramount importance here

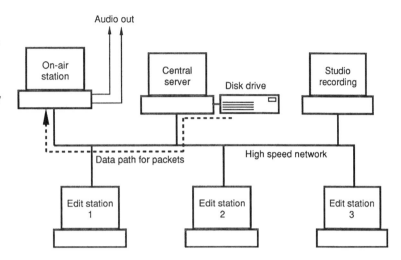

workstation, as illustrated in Figure 6.25. Here it must be possible to ensure guaranteed access to the remote disk at a rate adequate for real-time transfer, otherwise gaps will be heard in the replayed audio.

6.6.5 Audio networks

Recent networking solutions in the audio field, such as those proposed by AMS/Neve, Siemens, Sonic Solutions, Avid and others, use high speed optical fibre technology and FDDI or ATM. Other companies have stuck with the slower Ethernet for the time being, although not attempting to use it for the purpose of transferring large numbers of channels of audio in real time. Although Ethernet is capable of a maximum data rate of 10 Mbit/s, its performance under heavy load with many users is often as low as 2–3 Mbit/s because of the network access method used.

Sonic Solutions, for example, has taken the step of developing its own network protocol called SonicNet for transporting audio and other multimedia data over FDDI in a form which ensures optimum use of bandwidth by using large packets, and allows systems to guarantee that real-time file transfers, once started, will be able to complete without breaks (the so-called 'reservationist' approach).[14] It allows over 80 channels of audio to be transferred simultaneously. Phase 2 of SonicNet will allow the network to be used for routing real-time digital audio as well as for transferring files, thereby further blurring the distinction between digital audio interfaces and computer data networks.

211

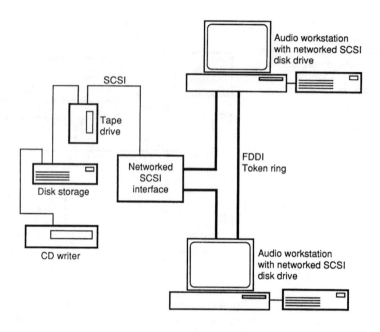

Figure 6.26 Example of a networked audio system with distributed storage and processing resources, allowing access to all disk and tape devices by both workstations (based on Sonic Solutions *SonicNet*)

Sonic Solutions has gone further through the OSI model than the first three layers specified in the FDDI protocol, allowing greater control over the way in which data is transferred. The alternative to this highly controlled approach is a form of network 'free for all', which assumes there is more network bandwidth than will ever be needed at any one time, allowing applications to grab whatever bandwidth they can for file transfer operations. This is simpler, but less reliable in cases of heavy network load.

An example of a networked audio system is shown in Figure 6.26, showing how resources can be distributed around the network. Note that this is based on SonicNet which allows stand alone disk drives to be connected directly to the network using a special interface chassis and card.

ATM allows for either guaranteed bandwidth communications between a source and a destination (needed for AV applications), or for more conventional variable bandwidth communication. It is a switched network, which means that the physical network is made up of a series of interconnected switches that are reconfigured to pass the information from sources to destinations according to the header information attached to each data packet. A network management system handles the negotiation between different devices that are contending for bandwidth, according to current demand.

6.6.6 Extending a network

It is common to need to extend a network to a wider area or to more machines. As the number of devices increases so does the traffic, and there comes a point when it is necessary to divide a network into zones, separated by 'repeaters', 'bridges' or 'routers'. Some of these devices allow network traffic to be contained within zones, only communicating between the zones when necessary. This is vital in large interconnected networks because otherwise data placed anywhere on the network would be present at every other point on the network, and overload could quickly occur.

A repeater is a device which links two separate segments of a network so that they can talk to each other, whereas a bridge isolates the two segments in normal use, only transferring data across the bridge when it has a destination address on the other

Figure 6.27 A WAN is formed by linking two or more LANs using a router

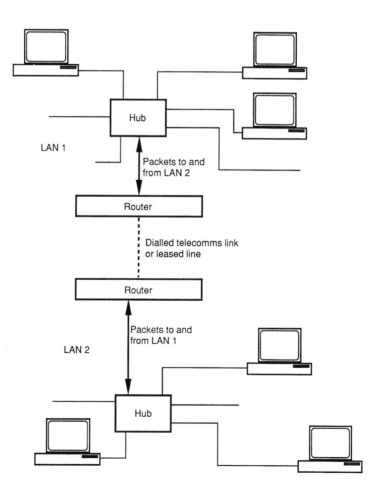

side. A router is very selective in that it examines data packets and decides whether or not to pass them depending on a number of factors. A router can be programmed only to pass certain protocols and only certain source and destination addresses. It therefore acts as something of a network policeman and can be used as a first level of ensuring security of a network from unwanted external access. Routers can also operate between different standards of network, such as between FDDI and Ethernet, and ensure that packets of data are transferred over the most time/cost-effective route.

One could also use some form of router to link a local network to another which was quite some distance away, forming a so-called Wide Area Network (WAN), as shown in Figure 6.27. Data can be routed either over dialled data links such as ISDN (see below), in which the time is charged according to usage just like a telephone call, or over leased digital circuits which are rented for a certain period of time from telecomms companies and are always available for data transfer without dialling. The choice would depend on the degree of usage and the relative costs. The data rate handled by such links is usually lower than that provided on the LAN, thereby limiting their usefulness for multichannel audio work, but still making them very useful for transferring files to distant destinations and for auditioning stereo audio material, possibly using a form of data reduction (see Chapter 3). Leased lines are often available at bandwidths of between 1 and 2 Mbit/s, whereas ISDN runs at 64 or 128 kbit/s. It is to be expected that lines with greater bandwidth will gradually become available at reasonable cost as time goes by.

6.6.7 SCSI assignment matrix

Although not a true network, it is possible to create some of the advantages of a network using a method of SCSI assignment, which allows central storage resources to be accessed from a variety of workstations. This is the approach adopted by SSL in its *SoundNet* system. In SoundNet up to 16 SCSI devices can be connected to a switcher which can assign the devices to different workstations, as shown in Figure 6.28. It requires that the workstation processors be located close to the switcher, because SCSI cables have to be short, so a lower speed Ethernet connection is then used to link the workstations for control purposes. The user terminals are then connected using RS422 serial interfaces.

This approach restricts the storage device assignment to a single machine at a time, but storage devices may be reassigned at any

Figure 6.28 Storage resources assigned to workstations using a SCSI assignment matrix (based on SSL *SoundNet*)

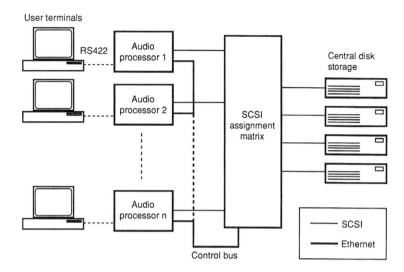

time so that a file could be quickly copied between two systems, for example. The Ethernet network can also be extended to the control terminal where an optical disk drive can be located for playing library disks. Ethernet is generally fast enough for transferring at least stereo audio in real time, provided not too many people try to do it at the same time. Multiple workstation processors can be synchronised together for large-scale projects involving multiple channels, each replaying eight channels of audio from separate disks.

6.6.8 ISDN

ISDN, the Integrated Services Digital Network, is a development in the telecommunications field which is gaining wider usage. ISDN is an extension of the digital telephone network to the consumer, providing two 64 kbit/s digital channels ('B' channels) which can be connected to ISDN terminals anywhere in the world simply by dialling. Data of virtually any kind may be transferred over the dialled-up link, and potential applications for ISDN include audio transfer.

Since the total usable capacity of a single ISDN-2 connection is only 128 kbit/s it is not possible to carry linear PCM data at normal audio resolutions over such a link, but it is possible to carry moderately high quality stereo digital audio at this rate using a data reduction system such as MPEG-Audio (see Chapter 3), or to achieve higher rates by combining more than one ISDN link to obtain data rates of, say, 256 or 384 kbit/s.[15] Multiple ISDN lines must be synchronised together using devices known as inverse multiplexers if the different delays which may arise over different connections are to be compensated. There are also ISDN-30 lines, providing 30 simultaneous 'B' channels of 64 kbit/s (giving roughly 2 Mbit/s), but these are expensive.

It is possible to use ISDN links for non-real-time audio file transfer, and this can be economically viable depending on the importance of the project and the size of files. The cost of an ISDN call is exactly the same as the equivalent duration of normal telephone call, and therefore it can be quite a cheap way of getting information from one place to another.

In the USA, there still remain a lot of circuits which are very similar to ISDN but not identical. These are called 'Switched 56' and carry data at 56 kbit/s rather than 64 kbit/s (the remaining 8 kbit/s that make up the total of 64 kbit/s is used for housekeeping data in Switched 56, whereas with ISDN the housekeeping data is transferred in a 'D' channel on top of the two 64 kbit data channels). This can create some problems when trying to link ISDN terminals, if there is a Switched 56 bridge somewhere in the way, requiring file transfer to take place at the lower rate.

For more information on ISDN, the reader is referred to an excellent book by John Griffiths.[16]

6.6.9 The Internet

There is now wide interest in the Internet as a means for worldwide communication. At the moment its primary use is not for real-time multimedia information such as audio and video, but there is growing evidence of applications which allow low quality (very low bit rate) audio and video information to be carried in this way. The RealAudio format, for example, developed by Progressive Networks, currently allows 8 bit, 11 kHz sampling rate, mono sound to be transferred in real time over the Internet, giving roughly medium wave radio quality, and with no guarantee of unbroken transfer. People are using the Internet for transferring some sound files in non-real time, but mainly short sounds for the games and music markets because of the limited data rates available to most users. Whole books have been written about the Internet, because it is a very large subject, so it is only proposed to give a brief introduction here.

The Internet is a collection of interlinked networks with bridges and routers in various locations, which originally developed amongst the academic and research community. It is used widely in these fields, but has also become extremely popular for personal and business applications, with many commercial operations now offering to connect users to the Internet for a reasonable fee. The bandwidth (data rate) available on the Internet varies from place to place, and depends on the route over which data is transferred. In this sense there is no easy way to guarantee a certain bandwidth, nor a certain 'time slot', and when there is a lot of traffic it simply takes a long time for data transfers to take place. Many people access the Internet through a commercial service provider, using a telephone line and a modem, and the maximum speed that is possible using such technology is currently 28.8 kbit/s (slow by audio standards). At this rate the error rate can be quite high, and many people operate at lower rates such as 2400, 9600 or 14400 bit/s. ISDN is another alternative for higher speed communications, and the most intensive users will probably opt for high speed leased lines giving direct access to the Internet. The inter-university network in the UK for example, called Super JANET, runs on leased lines at 10 Mbit/s.

The common protocol for communication on the Internet is called TCP/IP (Transmission Control Protocol/Internet Protocol). At a more detailed level, as part of the TCP/IP structure, there are high level protocols for transferring data in different ways. There is a file transfer protocol (FTP) used for downloading files from remote sites, a simple mail transfer protocol (SMTP) and a post office protocol (POP) for transferring email, and a hypertext transfer protocol (HTTP) used for interlinking sites on the so-called WorldWide Web (WWW). The WWW is a collection of file servers connected to the Internet, each with its own unique IP address (the method by which devices connected to the Internet are identified), upon which may be stored text, graphics, sounds and other data. Applications are available which make it easy to access these WWW servers, with links to applications that can replay any downloaded multimedia information.

References

1   Digidesign (1985, etc.) *Sound Designer I and II file format specifications*. Digidesign, 1360 Willow Rd, Menlo Park, CA 94025, USA. Tel: +1 415 688 0744; Fax: +1 415 327 3131.
2   Apple Computer (1991) *Audio Interchange File Format AIFF-C*. Apple Computer Inc., Developer Technical Support, 20525 Mariani Ave, MS:75-3T, Cupertino, CA 95014, USA. (Also available as a PostScript file by anonymous ftp from *ftp.sgi.com* in */sgi/aiff-c.9.26.91.ps*).

3   Electronic Arts (1985) *EA IFF 85 standard for interchange format files.*

4   Digigram (1993) *MusiFile format specification.* Digigram, Parc de Pré Milliet, 38330 Montbonnot, France. Tel: +33 76 52 47 47; Fax: +33 76 52 18 44.

5   Harris, J. and Ruben, I. (1992) *Bento Specification, Revision 1.0d4.1.* Apple Computer Inc.

6   Avid (1994) *OMF Interchange Specification Version 1.0.* OMF Developers' Desk, Avid Technology Inc., Metropolitan Technology Park, One Park West, Tewksbury, MA 01876, USA. Tel: +1 508 640 3400; Fax: +1 508 640 9768. Email: *omf-request@avid.com* .

7   Anon (1990) *DDP Reference Manual.* Doug Carson and Associates, 300 North Harrison, PO Box 1646, Cushing, OK 74023, USA. Tel: +1 918 225 0346; Fax: +1 918 225 1113.

8   Rumsey, F. and Watkinson, J. (1994) *The Digital Interface Handbook,* 2nd edition, Focal Press.

9   AES (1992) AES3-1992. 'Serial transmission format for two channel linearly represented digital audio data'. *Journal of the Audio Engineering Society* , **40**, 3, March

10   IEC (1992) *IEC 1119 Part 6: Serial Copy Management System.* International Electrotechnical Commission, Geneva.

11   AES (1991) AES10-1991 (ANSI S4.43-1991). 'Serial multichannel audio digital interface (MADI)'. *Journal of the Audio Engineering Society*, **39**, pp. 369–377.

12   AES (1991) AES11-1991 (ANSI 4.44-1991). *Synchronization of digital audio equipment in studio operations.* Audio Engineering Society.

13   ANSI (1990) ANSI X3.166-1990 *Physical Layer Medium Dependent.* American National Standards Institute.

14   Anderson, D. (1993) 'High speed networking for professional digital audio'. AES UK Digital Audio Interchange Conference, 18–19 May, London, pp. 60–69. Audio Engineering Society.

15   Burkhardtsmaier, B. *et al.* (1992) 'The ISDN MusicTAXI'. Presented at the 92nd AES Convention, Vienna, Austria, 24–27 March, preprint 3344.

16   Griffiths, J. (1990) *ISDN explained.* Wiley & Sons.

7 Video in the audio workstation

Video is now widely used alongside audio, and many workstations provide facilities for storing and replaying random-access or 'non-linear' video on disk. This has many uses, from low quality cue pictures to full-blown professional editing, and almost certainly uses some form of video data reduction to reduce the data rate of the picture to a point where it is suitable for storage on ordinary SCSI disks. Most commonly the JPEG compression process is used, which is available on third party processor boards for desktop multi-media systems, although the more recent MPEG may be used for some applications in the future. Some multimedia computers use proprietary video data reduction algorithms to store pictures at very low rates, making it possible to replay moderate quality moving images on ordinary computers. Apple's QuickTime is an example of this, although QuickTime can also use MPEG and JPEG if required.

People working with digital audio now need to have a basic understanding of digital video. This chapter will provide an introduction to the principles of digital video and video data reduction, as well as describing the SMPTE/EBU timecode format for synchronising video equipment with other sources. It will also examine some ways in which video capability can be added to digital audio workstations for post-production applications.

7.1 Digital video basics

7.1.1 A video picture

A video picture, such as might be viewed on a domestic television, is made up of horizontal lines, created by a spot scanning the screen from top to bottom. Each screen full of lines represents a still frame, and television systems often use a method of interlacing two sets of lines (each called a field) to make up one frame, presenting one field with gaps between the lines followed by the other filling in the missing lines to reduce flicker (see Figure 7.1). The field rate is thus twice the frame rate. There are 625 lines per frame in current European systems (PAL and SECAM) and 525 lines in American and Japanese systems (NTSC). The *frame rate* is the number of complete still frames per second making up the moving picture, being 25 Hz in Europe and approximately 29.97 Hz in countries such as America and Japan. This is called 'conventional definition television' or CDTV. In computer systems, images may be scanned and stored at a number of different rates and definitions. High definition television (HDTV) doubles the number of lines in the picture, amongst other things, in order to increase the resolution.

In audio, the term frequency relates to the pitch of a sound, being the repetition rate of the waveform. The human ear is most sensitive to middle frequencies around the 3–4 kHz region, and becomes increasingly less sensitive towards both extremes of the spectrum. If the high frequency response of an audio signal is reduced it becomes less 'bright' sounding, more 'muffled'. In video, the term frequency relates to the rate of change of visual information, but there are two dimensions in a still frame since the picture may be considered as an array of points or pixels (see Figure 7.2) in the horizontal and vertical dimensions, giving rise

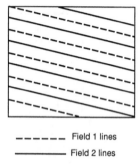

— — — — — Field 1 lines
————— Field 2 lines

Figure 7.1 Example of interlaced video scanning

Figure 7.2 A picture can be considered as an array of pixels or picture elements with a horizontal and vertical dimension. Spatial frequency refers to the rate of change of an image in either direction

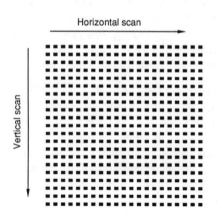

Horizontal scan

Vertical scan

Figure 7.3 Image a has a higher horizontal frequency than Image b

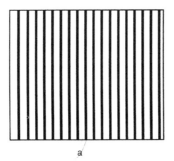
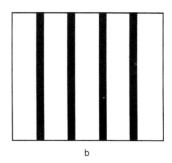

a   b

to the concept of 'spatial frequency'. There is also the concept of 'temporal frequency' which relates to the way in which the image changes over successive frames, as time passes. A high spatial frequency may be imagined as the fine detail in a picture, where picture elements change from light to dark in close succession, whereas a low frequency may be considered as a coarser pattern (see Figure 7.3). The human eye is less sensitive to high frequency visual information than to low frequencies, and it is less sensitive to the colour components of a signal than to its luminance or brightness components. The bandwidth of a typical CDTV picture signal in the analog domain is roughly 5.5 MHz.

In analog video, the horizontal dimension (a line) may be considered as a continuum of graduated shades from light to dark (in the monochrome sense), but the vertical dimension is effectively sampled (although not quantised) by the line structure. Digital video superimposes a sampling pattern on the horizontal dimension also, allocating a binary value to each pixel in the matrix, depending on its brightness. To represent colour, three video signals are normally required – red, green and blue (RGB video) – from which other colours and white may be formulated, but these are often combined in such a way as to produce a luminance (brightness) signal (Y) made up of an appropriate combination of R, G and B, and two 'colour difference' signals, namely R–Y (C_r) and B–Y (C_b), from which R, G and B may be reconstituted by matrixing them with Y. C_r and C_b can get away with a considerably lower bandwidth than Y, because the eye is much less sensitive to reduced colour bandwidth than it is to reduced luminance bandwidth. This colour difference format is often called 'component video' when referring to recording and transmission systems.

The term 'composite video' refers to a system in which the colour difference components have been modulated onto a subcarrier and combined with the luminance signal, creating one

221

waveform in which the colour information is spectrally inter-leaved with the luminance. The most common techniques for this purpose are PAL (Phase Alternate Line), used for 625/50 systems, and NTSC (National Television Standards Committee), used for 525/60. This is the most bandwidth-efficient form for video, with component video and RGB video requiring increasingly greater bandwidth. The process of modulation, though, results in some side effects due to interaction between the colour and monochrome information.

7.1.2 Digitising video

The CCIR 601 standard determined a basic sampling frequency for component video of 3.375 MHz which can be used in various multiples for luminance and colour signals. The so-called 4:2:2 format for video sampling uses 4 times 3.375 MHz (13.5 MHz) for the luminance component, and twice it for the colour difference signals (reduced bandwidth). The same sampling frequency is used for both 525/59.94 Hz TV systems as for 625/50 Hz formats, since it is a multiple of the line frequency of both, thus simplifying the design of equipment designed to work with both formats. Composite video is normally sampled at a different rate of 4 times the colour subcarrier frequency, thus resulting in different rates for 525 and 625 line video. This is due to the need to decode the colour information digitally, and simplifies the design of the required filters.

Typically, digital video samples are quantised to either 8 or 10 bit resolution – considerably lower than the 16 to 20 bit resolution used in audio. This is because the signal-to-noise ratio required for high video quality is not as great as that required for audio. The bit rate which therefore results from a 4:2:2 component digital video signal of either standard, sampled to 10 bit accuracy, is 270 Mbit/s. For composite video it is rather lower: e.g. 152 Mbit/s for 625/50 Hz. For HDTV, the bit rate for studio quality pictures is over 1000 Mbit/s (1 Gbit/s). Thus the data rate of uncompressed digital video exceeds that of digital audio by between 200 and over 1000 times.

7.2 Video data reduction

Video data reduction is important for many of the same reasons as audio data reduction (see Chapter 3): greater economy in storage and transmission, efficient use of bandwidth for broad-casting, and ease of use with consumer media and personal computers. In fact many of the same principles apply to video as to audio, since data reduction relies on the removal of redundant information from the signal, aiming to reduce the data rate with as little perceived effect on quality as possible. Many of the concepts introduced for audio data reduction can be transferred to the subject of video.

7.2.1 Introduction

As with audio data reduction, video data reduction is a business where virtually any compression factor is possible – but it rather depends on what final quality you are prepared to put up with.

To give a few examples: using MPEG-1 BRR (bit rate reduction) it is now possible to represent what the interactive media industry calls full motion video (FMV) at a bit rate as low as 1.2 Mbit/s, which is very similar to the rate required for full resolution audio. This represents a compression factor of over 100 times from the full composite data rate for PAL TV, and it is used to store conventional definition video onto CD-I for domestic distribution. The picture quality is only passable, but it stands comparison with VHS video recordings. High quality pictures can be achieved using MPEG-2 processing at a rate of around 5 Mbit/s.

It has also been shown that HDTV signals may be reduced in bit rate to around 20 Mbit/s or lower, with acceptably high picture quality for the majority of viewers, and tests have shown that it would be possible to carry digital HDTV in a conventional 8 MHz broadcast channel (as used for current analog TV broadcasts) using a suitable modulation method, which is indeed an achievement.

Figure 7.4 Block diagram of a simple video BRR encoder

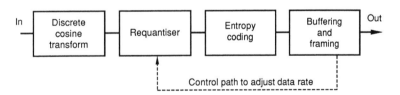

A block diagram of a simple coder for reducing the bit rate of digital video is shown in Figure 7.4. It shows only some of the techniques described below, and indicates the possibility for introducing some control feedback to the quantiser in order to keep the output bit rate constant.

223

Figure 7.5 Pixels are grouped into blocks and then subjected to a DCT to obtain spatial frequency components

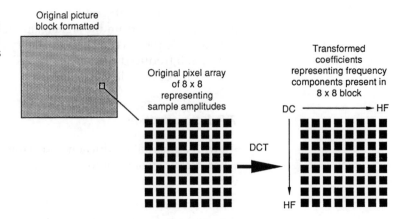

Original picture block formatted

Original pixel array of 8 x 8 representing sample amplitudes

Transformed coefficients representing frequency components present in 8 x 8 block

DC ⟶ HF

DCT

HF

7.2.2 Frequency domain transforms

Most approaches to video BRR divide the picture up into groups of pixels, changing it from a line structure into a block structure (see Figure 7.5). A common block size is 8 x 8 pixels, although systems using other possibilities such as 16 x 16 or 8 x 4 are in evidence. This block structure may often be seen in still pictures which have been data reduced if they are magnified considerably, or if a very high compression factor has been used. The sample values corresponding to the pixels represent their amplitudes (or brightness in a monochrome picture). The block is then subjected to a discrete cosine transform (DCT) as in many audio BRR systems, which results in a frequency domain representation of the block, whose sample values now represent the relative strengths of different spatial frequencies in the block, both horizontal and vertical. This is an efficient approach to BRR in itself since many of the spatial frequency components will be near zero in the typical picture block, with only a few of significant amplitude which will be stored or transmitted.

To give an example, a block with a finely detailed black and white pattern would result in a high frequency component at the pattern repetition frequency after the DCT process. Most other components would be zero. Applying an inverse DCT would restore the original spatial amplitudes of the pixels. If one were to remove the zero or near zero samples in the frequency domain, then perform the inverse DCT, the reconstituted picture would probably look almost identical to the original. Even without any requantisation this technique can result in a significant reduction in the amount of information required to represent a block.

7.2.3 Psychovisual model

Since the eye is less sensitive to high frequency information and to colour information it is possible to allow more noise in both these regions than in the low frequency luminance information. Thus the spectral coefficients produced by the DCT may be weighted and requantised according to a table which is based on the visibility of noise. These weighting and quantisation parameters may be varied to control the average bit rate of the audio signal, which would otherwise vary considerably from block to block.

7.2.4 Entropy coding

The spectral coefficients, thus requantised, may be subjected to a variable length or Huffmann coding process which assigns a binary code of a certain length to each sample. Based on an analysis of which bit patterns occur most often in typical TV pictures it is possible to assign the shortest codes to the most common patterns, and the longest to the least common, thus introducing a further saving in data rate.

7.2.5 Inter- and intra-frame coding

In addition to the aforementioned methods, it is also possible to achieve even greater reductions in the bit rate by analysing the temporal aspects of the moving picture, attempting to predict how the picture changes over a number of frames. This is called interframe coding. When coding is simply performed within the field or frame itself, the process is called intrafield or -frame coding. Which of these methods is appropriate depends on the type of data reduction process in use, the application, and often there is some analysis of which will provide the greatest reduction in bit rate after variable length coding, allowing the system to switch to the most suitable 'on the fly'.

The interframe mode relies on attempting to predict the sample values in the block, based on the same block or a motion-compensated version of it in the previous frame, performing the DCT only on the error or difference between the predicted and the actual sample values. Basing the prediction on the same (co-sited) block in the previous frame will only be useful if the picture is static. If it is moving then it will be necessary to predict the direction in which a block (representing a picture element) is moving, by making searches in each direction and comparing the current frame to the previous frame, attempting to find the best match with the current frame's block. A displacement vector can then be determined, and transmitted along with the coded picture coefficients, so that it may be used by the decoder to form a similar motion-compensated prediction.

Such motion compensated inter-field coding requires that a simple intrafield mode is used every so often for each block, in order that prediction errors do not accumulate. In some computer-based systems, such as Apple's QuickTime video compressor, a similar approach is used, called 'frame differencing', whereby only the differences between successive frames are coded, it being possible to set the distance between 'key frames' which are full frames rather than difference coefficients. The less frequently these 'key frames' are stored, the poorer the picture quality. Interframe coding does not always result in savings in the bit rate, since not all images move in a predictable way, and backgrounds may be confusing, and in such cases intraframe coding is more appropriate.

7.2.6 Skip field

A technique also used in some non-linear video editors involves coding only every other field of the picture, repeating the first field with a small vertical shift to create the second field of each frame on replay. This clearly halves the amount of data to be stored but also halves the vertical resolution of the picture. It can be adequate for deriving a picture for off-line editing, but would not be used for broadcast quality video.

7.2.7 Buffering

All of these techniques applied to a typical video signal will tend to result in a bit rate which varies considerably with time unless something is done to keep it constant. In computer still-picture storage applications it may not matter that the disk space taken up by one picture is different from another, but in most transmission and storage applications it is necessary to have a constant bit rate. This may be achieved by buffering the output of the coder using memory from which data is clocked out at a constant rate. The fullness of the buffer can be used to control the requantiser or coefficient weighting earlier in the chain, adjusting the accuracy of requantisation of the DCT coefficients to suit the required bit rate at that instant. Alternatively a form of forward prediction may be used to estimate the entropy of the final coded signal, and this used to control the quantiser to maintain a constant data rate.

7.3 MPEG and JPEG

JPEG was intended originally for high quality still frame pictures, and is used widely in computer graphics and multimedia applications for compressing 24-bit still images. It is possible to trade off compression factor against picture quality, and the compression process can either be carried out relatively slowly using an off-the-shelf software package, or quickly using dedicated hardware such as the C-Cube JPEG compression chip, which is fast enough to compress images at FMV frame rates for computer applications. Motion JPEG is basically still-frame JPEG adapted for moving pictures, but based on the same process. Motion JPEG tends to be used in studio applications where editing and other post-processing is required, because JPEG frames are complete within themselves (interframe coding) and a recording can be edited anywhere. JPEG processing allows good still frame and other trick-mode performance.

MPEG-style coding may involve prediction and frame difference techniques such as those described above, and thus is most appropriate for transmission or replay of video programme material where the picture sequence is presented in a time continuum and in the forward direction, as opposed to studio recording or storage where slow motion, shuttle, reverse and still frame modes may be required. MPEG-1[1] was optimised for non-interlaced pictures, whereas the more recent MPEG-2[2] can code CDTV pictures at rates from about 4–9 Mbit/s and HDTV pictures at rates from 15–25 Mbit/s, and can deal with interlaced pictures. Table 7.1 shows an indication of the bit rates required for MPEG-2 to give various picture qualities. Because of the greater degree of data reduction obtainable from MPEG compared with JPEG, there is now considerable interest in developing non-linear editing systems based on MPEG-2, using various ingenious processes to circumvent the limitations of MPEG as a studio recording format.[3]

Table 7.1 MPEG-2 data rate versus picture quality

Data rate (Mbit/s)	Equivalent quality
4–5	Domestic conventional TV pictures
8–10	Studio quality conventional TV pictures
16–20	Domestic HDTV pictures
32–40	Studio HDTV pictures

7.4 Desktop video

Apple's QuickTime video is an example of technology used to add video recording and replay capabilities to desktop computers. It is part of the company's line up of multimedia software, designed for recording and reproducing moving pictures on all but the most lowly computers. Originally only available for the Mac, it has now been ported for the PC as well. It started out as very basic video, running at 15 frames per second and viewable in a small window on the computer screen, compressed and decompressed using a proprietary Apple algorithm (although other options were available). QuickTime movies could be stored in relatively little disk space, and used a crude means of maintaining synchronisation with any accompanying audio. Acknowledging that audio replay should be continuous, and that QuickTime movies had to be able to replay on virtually any machine, fast or slow, synchronisation was maintained by dropping video frames when the storage device was unable to transfer data fast enough. Movies played back on slow machines in a jerky fashion, and on faster machines they looked better and played more smoothly.

Recently QuickTime has been extended to offer full-frame video at normal TV frame rates. It also encompasses MPEG video, which will have the effect of allowing MPEG-encoded movies to be replayed on desktop computers. JPEG boards can also be used in conjunction with desktop computers to give high quality movie replay with straightforward still frame handling. QuickTime 2 will also allow timecode to be recorded as a track alongside audio and video. Such technology is becoming widely used to provide video cue pictures for audio workstations, for sound dubbing and other post-production applications, with some companies enhancing QuickTime to provide better synchronisation between sound and picture. Digidesign's *PostView* is an example of such a package.

The Microsoft equivalent of QuickTime is Video for Windows. This uses similar processes to allow video capability to be added to MS-DOS PCs running Windows. It forms part of the Microsoft MPC standard for multimedia PCs. MPEG video replay is also an integral part of the most recent MPC standard.

7.5 Digital video options for workstations

Adding a digital video option to an audio workstation will allow pictures to be stored on disk in the same way as audio, with the same advantage of random access. Prior to this, those wishing to dub sound to picture had to synchronise the workstation to an external video tape recorder, using serial remote control from the workstation. This normally resulted in considerable delays whilst the VTR transport located, parked and played, and also might require a few seconds for the workstation to lock up to the video timecode. For repeating short loops and other such situations the time wasted was considerable. Random access video removes most of these problems by allowing any section of the picture to be located immediately, and for playing, nudging and rewinding to be quickly locked to the audio. The only slight drawback is the same as for audio – the need to load the video on to the hard disk before starting.

The most common way of adding video capability to a desktop audio workstation based on a Mac or PC is to purchase one of the many third party video capture and compression boards on the market. This assumes that the computer does not already have video recording and replay capability already on board, as some recent multimedia machines do. Even if the computer has video hardware installed it is likely that it will be fairly basic, and may not operate using MPEG or JPEG compression. JPEG boards are currently the most widely used, and allow real-time video to be recorded and replayed at full-frame size, at up to 30 frames per second. It is likely that MPEG-2 will also become popular, although see above for its limitations and advantages. The quality of these boards varies, and so does the software that goes with them. If high quality pictures are required, with good slow motion and other trick modes, it pays to buy a quality product.

It is often recommended that a separate fast hard disk is used for storing the compressed video, in order that a sufficiently high data rate can be obtained for smooth picture replay. Usually the software allows the video data rate to be adjusted by the user in order to trade off storage space and transfer rate against picture quality and size. SCSI-2 interfaces to disk drives are often recommended by manufacturers, again because of the need for reasonably high transfer rates, and RAID systems are also considered to be a good thing for this purpose (see Chapter 4).

The video hardware and software is not the only thing you will need. Many audio manufacturers now sell software to link audio and video elements of the package, in order to ensure that the two remain in synchronism, and that the control of the two can be carried out from the same interface. It is normally possible to decide which of sound and picture are master and slave. As with all such products, it is important to ensure that the hardware and software you choose all works together, since there are many incompatible options on the market. Since manufacturers change their products on an almost monthly basis it would be impossible to give any specific product-related guidance here.

Some purpose-designed audio workstations (those not based around a desktop computer) also offer digital video facilities, based on integral hardware and software, sometimes using proprietary compression algorithms. SSL's VisionTrack system for its *Scenaria* and *Omnimix* products is an example of such an approach. This includes software capable of displaying a graphical timeline of the picture information, showing where key changes in the picture have occurred, making it easier to locate picture cut points quickly.

7.6 SMPTE/EBU timecode

The American Society of Motion Picture and Television Engineers (SMPTE) proposed a system to facilitate the accurate editing of video tape in 1967. This became known as SMPTE ('simpty') code, and it is basically a continuously running eight-digit clock registering time from an arbitrary start point (which may be the time of day) in hours, minutes, seconds and frames, against which the programme runs. The clock information is encoded into a signal which can be recorded as an audio signal (LTC, or longitudinal timecode) or inserted into lines at the top of the television picture (VITC, or vertical interval timecode). Every single frame of a video recording has its own unique number called the timecode address and this can be used to pinpoint a precise editing position.

A number of frame rates are used, depending on the television standard to which they relate: 30 frames per second (fps), or true SMPTE, was used for monochrome American television, and is now only used for CD mastering in the Sony 1630 format; 29.97 fps is used for colour NTSC television (mainly USA, Japan and parts of the Middle East), and is called 'SMPTE drop-frame' (see below); 25 fps is used for PAL and SECAM TV and is called 'EBU' (Europe, Australia, etc.); and 24 fps is used for some film work.

Each timecode frame is represented by an 80 bit binary 'word', split principally into groups of 4 bits, with each 4 bits representing a particular parameter such as tens of hours, units of hours, and so forth, in BCD (binary-coded decimal) form (see Figure 7.6). Sometimes, not all four bits per group are required — the hours only go up to '23', for example — and in these cases

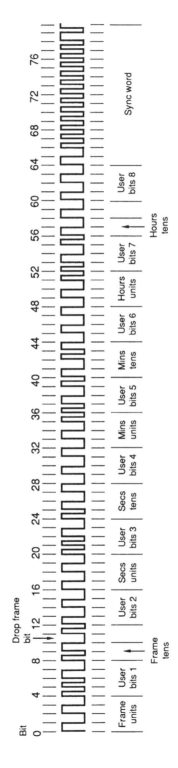

Figure 7.6 Data format of the SMPTE/EBU longitudinal timecode frame. Not the 'sync word' 0011111111111101 which occurs at the end of each frame to mark the boundary. This pattern does not occur elsewhere in the frame and its asymmetry allows a timecode reader to determine the direction in which the code is being played (forwards or backwards)

Figure 7.7 The FM or biphase-mark channel code is used to modulate the timecode data so that it can be recorded as an audio signal

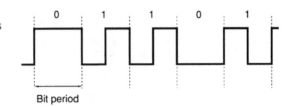

the remaining bits are either used for special control purposes or set to zero (unassigned): 26 bits in total are used for time address information to give each frame its unique hours, minutes, seconds, frame value; 32 are 'user bits' and can be used for encoding information such as reel number, scene number, day of the month and the like; bit 10 can denote drop-frame mode if a binary 1 is encoded there, and bit 11 can denote colour frame mode if a binary 1 is encoded. The end of each word consists of 16 bits in a unique sequence, called the 'sync word', and this is used to mark the boundary between one frame and the next. It also allows the reader to tell in which direction the code is being read, since the sync word begins with 11 in one direction and 10 in the other.

This binary information cannot be recorded directly, since its bandwidth would be too wide, so it is modulated in a simple scheme known as 'bi-phase mark', or FM, such that a transition from one state to the other (low to high or high to low) occurs at the edge of each bit period, but an additional transition is forced within the period to denote a binary 1 (see Figure 7.7). The result looks rather like a square wave with two frequencies, depending on the presence of ones and zeros in the code. Depending on the frame rate, the maximum frequency of square wave contained within the timecode signal is either 2400 Hz (80 bits x 30 fps) or 2000 Hz (80 bits x 25 fps), and the lowest frequency is either 1200 Hz or 1000 Hz, and thus it may easily be recorded on an audio machine. The code can be read forwards or backwards, and phase inverted. Readers are available which will read timecode over a very wide range of speeds, from around 0.1 to 200 times play speed. The rise-time of the signal, that is the time it takes to swing between its two extremes, is specified as 25 µs ± 5 µs, and this requires an audio bandwidth of about 10 kHz.

VITC is used widely in video recorders and is recorded not on an audio track, but in the lines within the vertical sync period of a video picture, such that it can always be read when video is capable of being read, such as in slow-motion and pause modes.

7.7 Drop-frame timecode

When colour TV (NTSC standard) was introduced in the USA it proved necessary to change the frame rate of monochrome TV slightly in order to accommodate the colour information within the same spectrum. The 30 fps of monochrome TV, originally chosen so as to lock to the American mains frequency of 60 Hz, was thus changed to 29.97 fps, since there was no longer a need to maintain synchronism with the mains owing to improvements in oscillator stability. In order that 30 fps timecode could be made synchronous with the new frame rate it was necessary to drop two frames at the start of every minute, except for every tenth minute, which resulted in minimal long-term drift between timecode and picture (75 ms over 24 hours). The drift in the short term gradually increased towards the minute boundaries and was then reset.

A flag is set in the timecode word to denote NTSC drop-frame timecode. This type of code should be used for all applications where the recording might be expected to lock to an NTSC video programme. Users and software must be aware of the lack of the first two frames of most minutes, and adapt accordingly.

7.8 Handling timecode

Timecode may be recorded or 'striped' on to audio and video tapes before, during or after the programme material is recorded, depending on the application. Normally the timecode must be locked to the same speed reference as that used to lock the speed of the tape machine, otherwise a long-term drift can build up between the passage of time on the tape and the measured passage in terms of timecode. The sync reference for a timecode generator is usually provided in the form of a video composite sync signal, and video sync inputs are increasingly provided on digital tape recorders for this purpose.

Timecode generators are available in a number of forms, either as stand-alone devices, as part of a synchroniser or editor, or integrally within a workstation. In large centres timecode is sometimes centrally distributed and available on a jackfield point. When generated externally, timecode normally appears as an audio signal on an XLR connector or jack. In MIDI systems it is common to use MIDI TimeCode (MTC), which is a version of SMPTE/EBU code carried over a MIDI interface (see section 8.10), and quite a few desktop workstations incorporating audio and a MIDI sequencer use MTC as the sync source. Most generators allow the user to preset the start time and the frame-rate

233

standard.

Timecode is often recorded on to an outside track of a multitrack tape machine (usually track 24), or a separate timecode or cue track will be provided on digital machines. The signal is recorded at around 10 dB below reference level, and crosstalk between tracks or cables is often a problem due to the very audible mid-frequency nature of timecode. Professional DAT machines are often capable of recording timecode, this being converted internally into a DAT running-time code which is recorded in the subcode area of the digital recording. On replay, any frame rate of timecode can be derived, no matter what was used during recording, which is useful in mixed-standard environments. On video tape recorders, LTC is normally recorded either on a dedicated timecode track (professional machines), on a spare audio track, or additionally in the picture's vertical sync interval (VITC).

In mobile film and video work which often employs separate machines for recording sound and picture it is necessary to stripe timecode on both the camera's tape or film and on the audio tape. This can be done by using the same timecode generator to feed both machines, but more usually each machine will carry its own generator and the clocks will be synchronised at the beginning of each day's shooting, both reading absolute time of day. Highly stable crystal control ensures that sync between the clocks will be maintained throughout the day, and it does not then matter whether the two (or more) machines are run at different times or for different lengths of time because each frame has a unique time of day address code which enables successful post-production syncing.

When recording timecode on tapes the code should run for around 30 seconds or more before the programme begins in order to give other machines and computers time to lock in. If programme is spread over several reels, the timecode generator should be set and run such that no number repeats itself anywhere throughout the reels, thus avoiding confusion during post-production. Alternatively the reels can be separately numbered.

With digital audio workstations, timecode is not usually recorded as a signal, but is often used to reference the replay of digital audio files, using suitable clock ratio conversions to the audio sampling rate, as described in section 5.5. The original timecode start times and frame rates of sound files may be stored in the file header for subsequent reference.

References

1   International Standards Organisation (1993) *ISO/IEC 11172: Information Technology – Coding of moving pictures and associated audio for digital storage media at up to about 1.5 Mbit/s.*

2   International Standards Organisation (1994) *ISO/IEC 13818 – Generic coding of moving pictures and associated audio.* March.

3   Stone, J. (1995) 'Studio compression'. *IEE New Broadcast Standards and Systems*, 3–7 July, Durham University. Institute of Electrical Engineers, London.

8 MIDI, sampling and synthesis for workstations

MIDI, the Musical Instrument Digital Interface, now forms an integral part of many studio and multimedia operations. It can be used for controlling musical instruments such as samplers and synthesisers, internal sound cards within PCs (on which may be provided FM and wavetable synthesis functions, as well as audio sampling), and a range of other studio equipment such as mixers, effects devices and recording devices. MIDI software is often integrated with digital audio software, so as to produce a package capable of recording, editing and replaying both musical performance data and sampled digital audio data, combining the benefits of both approaches.

MIDI has become popular in multimedia workstations as a means of controlling and storing sound information in an economical way. As previously explained, sampled digital audio can occupy very large amounts of memory space (around 5 Mbytes per minute for high quality sound), whereas the MIDI data for a minute of music might only consume 5 kbytes. The overall processing load on a computer when replaying MIDI data is also very much lower than when replaying digital audio. Whilst the two are not directly comparable, the widespread adoption of General MIDI and the availability of cheap sound cards for PCs has meant that many high quality computer sound applications can be handled using MIDI control instead of digital sound files.

This chapter will provide an introduction to MIDI and its use in computer-based workstations. For further details the reader is referred to the book *MIDI Systems and Control* as detailed at the end of this chapter.

8.1 What is MIDI?

MIDI is a serial remote control interface first developed for musical systems. It is a measure of the popularity of MIDI as a means of control that it has now been adopted in many other audio and visual systems, including the automation of mixing consoles, the control of studio outboard equipment, the control of lighting equipment and of other studio machinery. Although many of its standard commands are music related, it is possible either to adapt music commands to non-musical purposes or to use command sequences designed especially for alternative methods of control. MIDI integrates timing and system control commands with pitch and note triggering commands. It is possible to control musical instruments polyphonically in pseudo real time: in other words, the speed of transmission is such that delays in the transfer of performance commands are not audible in the majority of cases. It is also possible to address a number of separate receiving devices within a single MIDI data stream.

The adoption of a serial standard for MIDI was dictated largely by economic and practical considerations, as it was intended that it should be possible for the interface to be installed on relatively cheap items of equipment, and that it should be available to as wide a range of users as possible. The simplicity and ease of installation of MIDI systems has been largely responsible for its rapid proliferation as an international standard.

8.2 MIDI and digital audio contrasted

For many the distinction between MIDI and digital audio may be a clear one, but those new to the subject often confuse the two. Any confusion is often due to both MIDI and digital audio equipment appearing to perform the same task – that is the recording of multiple channels of music using digital equipment – and is not helped by the way in which some manufacturers refer to MIDI sequencing as digital recording.

Digital audio involves a process whereby an audio waveform (such as the line output of a musical instrument) is sampled regularly and then converted into a series of binary words which actually represent the sound itself (see Chapter 2). A digital audio recorder stores this sequence of data and can replay it by passing the original data through a digital-to-analog convertor which turns the data back into a sound waveform, as shown in

Figure 8.1 (a) Digital audio recording and (b) MIDI recording contrasted. In (a) the sound waveform itself is converted into digital data and stored, whereas in (b) only control information is stored, and a MIDI-controlled sound generator is required during replay

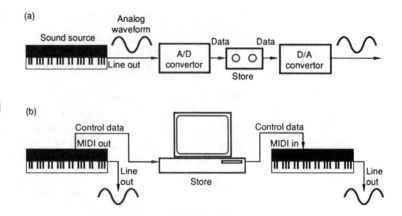

Figure 8.1. A multitrack recorder has a number of independent channels which work in the same way, allowing a sound recording to be built up in layers. MIDI, on the other hand, handles digital information which controls the generation of sound. MIDI data does not represent the sound waveform itself. When a multitrack music recording is made using a MIDI sequencer this control data is stored, and can be replayed by transmitting the original data to a collection of MIDI-controlled musical instruments. It is the instruments which actually reproduce the recording.

A digital audio recording, then, allows any sound to be stored and replayed without the need for additional hardware. It is useful for recording acoustic sounds such as voices, where MIDI is not much help. A MIDI recording is almost useless without a collection of sound generators. An interesting advantage of the MIDI recording is that, since the stored data represents the events in a piece of music, it is possible to change the music by changing the event data. MIDI recordings also consume a lot less memory space and require far lower transfer rates than digital audio recordings, and in this sense MIDI is an ideal form of data reduction! For this reason, software designers may prefer to generate sounds for games and other such applications using MIDI to control a synthesised or sampled sound card, rather than replaying real digital audio.

8.3 Basic MIDI principles

8.3.1 System specifications

The MIDI standard specifies a uni-directional serial interface running at 31.25 kbits ±1 per cent. The rate was defined at a time when the clock speeds of microprocessors were typically much

Figure 8.2 A MIDI message consists of a number of bytes, each transmitted serially and asynchronously by a UART in this format, with a start and stop bit to synchronise the receiving UART. The total period of a MIDI data byte, including start and stop bits, is 320 µs

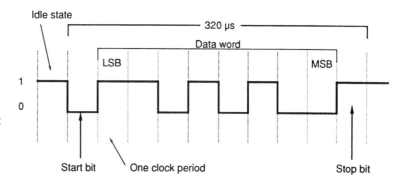

slower than they are today, this rate being a convenient division of the typical 1 or 2 MHz master clock rate. The rate had to be slow enough to be carried without excessive losses over simple cables and interface hardware, but fast enough to allow musical information to be transferred from one instrument to another without noticeable delays. Data is transmitted unidirectionally, that is from the transmitter to the receiver, and there is no return path unless a separate MIDI link is made.

Control messages are sent as groups of bytes, preceded by one start bit and followed by one stop bit per byte in order to synchronise reception of the data which is transmitted asynchronously, as shown in Figure 8.2. The addition of start and stop bits means that each 8 bit word actually takes ten bit periods to transmit (lasting a total of 320 µs) – a factor which must be borne in mind when attempting to calculate how long a particular message will take to transmit. Standard MIDI messages typically consist of one, two or three bytes, although there are longer messages for some purposes which will be covered later in this book.

There is a defined hardware interface which should be incorporated in all MIDI equipment, which ensures electrical compatibility within a system. This is shown in Figure 8.3. Most equipment using MIDI has three interface connectors: IN, OUT, and THRU (*sic*). The OUT connector carries data which the device itself has generated. The IN connector receives data from other devices and relays it to the device's UART (see Chapter 1). The THRU connector is a direct throughput of the data that is present at the IN. As can be seen from the hardware interface diagram, it is simply a buffered feed of the input data, and it has not been processed in any way. A few cheaper devices do not have THRU connectors, but it is possible to obtain 'MIDI THRU boxes' which provide a number of 'THRUs' from one input. Occasionally, devices without a THRU socket allow the

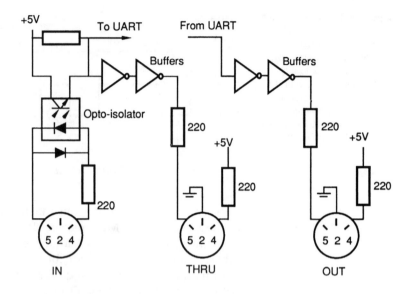

Figure 8.3 MIDI electrical interface showing IN, OUT and THRU ports

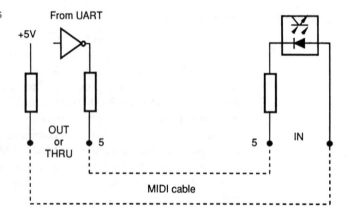

Figure 8.4 A current loop is formed between the OUT of the transmitter and the IN of the receiver when a MIDI cable is connected. The LED in the receiver's opto-isolator is turned on and off according to current flow

OUT socket to be switched between OUT and THRU functions. A 5 mA current loop is created between a MIDI OUT or THRU and a MIDI IN, when connected with the appropriate cable, and data bits are signalled by the turning on and off of this current by the sending device. This principle is shown in Figure 8.4.

The interface incorporates an opto-isolator between the MIDI IN (that is the receiving socket) and the device's microprocessor system. This is to ensure that there is no direct electrical link between devices, and helps to reduce the effects of any problems which might occur if one instrument in a system were to develop an electrical fault. An opto-isolator is an encapsulated device in which a light-emitting diode (LED) can be turned on or off

Figure 8.5 The edges of a square pulse subjected to rise-time distortion

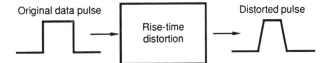

depending on the voltage applied across its terminals, illuminating a photo-transistor which consequently conducts or not, depending on the state of the LED. Thus the data is transferred optically, rather than electrically. In the MIDI specification, the opto-isolator is defined as having a rise time of no more than 2 µs. The rise time affects the speed with which the device reacts to a change in its input, and if slow will tend to distort the leading edge of data bit cells, as shown in Figure 8.5. The same also applies in practice to fall times.

Rise-time distortion results in timing instability of the data, since it alters the time at which a data edge crosses the decision point between one and zero, and if the rise time is excessively slow the data value may be corrupted since the output of the device will not have risen to its full value before the next data bit arrives. If a large number of MIDI devices are wired in series (that is from THRU to IN a number of times) the data will be forced to pass through a number of opto-isolators and thus will suffer the combined effects of a number of stages of rise-time distortion. Whether or not this will be sufficient to result in data detection errors at the final receiver will depend to some extent on the quality of the opto-isolators concerned, and also on other losses which the signal may have suffered on its travels. It follows that the better the specification of the opto-isolator, the more stages of device cascading will be possible before unacceptable distortion is introduced.

Concerning any potential for delay between IN and THRU connectors, it should be stated at the outset that the delay in data passed through this circuit is only a matter of microseconds, so this contributes little to any audible delays perceived in the musical outputs of some instruments in a large system. The bulk of any perceived delay will be due to other factors, which are covered in later sections.

8.3.2 Connectors and cables

The connectors used for MIDI interfaces are like the 5 pin DIN plugs used in some hi-fi systems, and although it is possible to use hi-fi cables (depending on the way that they are wired), better quality connectors are to be preferred. The specification also allows for the use of XLR-type connectors (such as those used for balanced audio signals in professional equipment), although these are rarely encountered in practice. Only three of the pins of a 5 pin DIN plug are actually used in most equipment (the three innermost pins).

The cable should be a shielded twisted pair with the shield connected to pin 2 of the connector at both ends, although within the receiver itself, as can be seen from the diagram above, the MIDI IN does not have pin 2 connected to earth. This is to avoid earth loops, and makes it possible to use a cable either way round. (If two devices are connected together whose earths are at slightly different potentials, a current is caused to flow down any earth wire connecting them. This can induce interference into the data wires, possibly corrupting the data, and can also result in interference such as hum on audio circuits.)

It is recommended that no more than 15 m of cable is used for a single cable run in a simple MIDI system, and investigation of typical cables indicates that corruption of data does indeed ensue after longer distances, although this is gradual and depends on the electromagnetic interference conditions, the quality of cable and the equipment in use. Longer distances may be accommodated with the use of buffer or 'booster' boxes which act to compensate for some of the cable losses and retransmit the data. It is also possible to extend a MIDI system by using a networking approach, with appropriate hardware to connect the MIDI cable to a high speed data network.

In the cable, pin 5 at one end should be connected to pin 5 at the other, and likewise pin 4 to pin 4, and pin 2 to pin 2. Unless any hi-fi DIN cables to be used follow this convention they will not work. Professional microphone cable terminated in DIN connectors may be used as a higher-quality solution, because domestic cables will not always be a shielded twisted-pair and thus are more susceptible to external interference, as well as radiating more themselves which could interfere with adjacent audio signals. It is recommended that the correct cable is used in professional installations where MIDI cables are installed in the same trunking as audio cables, to avoid any problems with crosstalk.

8.3.3 Simple interconnection

Before going further it is necessary to look at a simple practical application of MIDI, to see how various aspects of the standard message protocol have arisen. For the present it will be assumed that we are dealing with a music system, although, as will be seen, MIDI systems can easily incorporate non-musical devices in various ways.

The IN connector is used to control a MIDI device, whereas the OUT connector carries commands generated by the device, such as keys pressed, controls altered, and so forth. The THRU socket can be used to 'daisy-chain' MIDI devices together, so that transmitted information from one controller can be sent to a number of receivers without the need for multiple outputs from a controller. Occasionally devices may have an internal 'merging' function which merges data from the device's front panel with data from the MIDI IN, sending the combined data to the MIDI OUT.

Figure 8.6 The simplest form of MIDI interconnection involves connecting two instruments together as shown

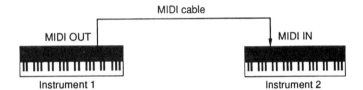

In the simplest MIDI system, one instrument could be connected to another as shown in Figure 8.6. Here, instrument 1 sends information relating to actions performed on its own controls (notes pressed, pedals pressed, etc.) to instrument 2, which imitates these actions as far as it is able. This type of arrangement can be used for 'doubling-up' sounds, 'layering' or 'stacking', such that a composite sound can be made up from two synthesisers' outputs. (It should be pointed out that the audio outputs of the two instruments would have to be mixed together for this effect to be heard.) Larger MIDI systems could be built up by further 'daisy-chaining' of instruments, such that instruments further down the chain all received information generated by the first (see Figure 8.7), although this is not a very satisfactory way of building a large MIDI system. In such cases it is more appropriate to use multiport MIDI interfaces, such as described later in this chapter.

Figure 8.7 Further instruments can be added using THRU ports as shown, in order that messages from instrument 1 may be transmitted to all the other instruments

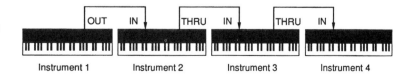

8.3.4 MIDI channels

MIDI messages are made up of a number of bytes. Each part of the message has a specific purpose, and one of these is to define the receiving channel to which the message refers. In this way, a controlling device can make data device-specific – in other words it can define which receiving instrument will act on the data sent. This is most important in large systems which use a computer sequencer as a master controller, when a large amount of information will be present on the MIDI data bus, not all of which is intended for every instrument. If a device is set in software to receive on a specific channel or on a number of channels it will act only on information which is 'tagged' with its own channel numbers. Everything else it will usually ignore. There are sixteen basic MIDI channels and instruments can usually be set to receive on any specific channel or channels (omni off mode), or to receive on all channels (omni on mode).

Later it will be seen that the limit of sixteen MIDI channels can be exceeded easily by using multiport MIDI interfaces connected to a computer. In such cases it is important not to confuse the MIDI data channel with the physical port to which a device may be connected, since each physical port will be capable of transmitting on all sixteen data channels.

8.3.5 Message format

There are two basic types of MIDI message byte: the status byte and the data byte. Status bytes always begin with a binary one to distinguish them from data bytes, which always begin with a zero. As shown in Figure 8.8, the first half of the status byte denotes the message type and the second half denotes the channel number. Because the most significant bit (MSB) of each byte is reserved to denote the type (status or data) there are only seven active bits per byte which allows 2^7 (that is 128) possible values.

Figure 8.8 General format of a MIDI message. The 'sss' bits are used to define the message type, the 'nnnn' bits define the channel number, whilst the 'xxxxxxx' and 'yyyyyyy' bits carry the message data. See text for details

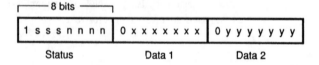

The first byte in a MIDI message is normally a status byte, which contains information about the channel number to which the message applies. It can be seen that four bits of the status byte

Table 8.1 MIDI messages summarised

Message	Status	Data 1	Data 2
Channel messages			
Note off	&8n	Note number	Velocity
Note on	&9n	Note number	Velocity
Polyphonic aftertouch	&An	Note number	Pressure
Control change	&Bn	Controller number	Data
Program change	&Cn	Program number	–
Channel aftertouch	&Dn	Pressure	–
Pitch wheel	&En	LSbyte	MSbyte
System exclusive			
System exclusive start	&F0	Manufacturer ID	Many data bytes
End of SysEx	&F7	–	
System common			
Quarter frame (MTC)	&F1	Data	–
Song pointer	&F2	LSbyte	MSbyte
Song select	&F3	Song number	–
Tune request	&F6	–	
System real-time			
Timing clock	&F8	–	–
Start	&FA	–	–
Continue	&FB	–	–
Stop	&FC	–	–
Active sensing	&FE	–	–
Reset	&FF	–	–

are set aside to indicate the channel number, which allows for 2^4 (or 16) possible channels. The status byte is the label that denotes which receiver the message is intended for, and it also denotes which type of message is to follow (e.g. a note on message). It will also be seen that there are three bits to denote the message type, because the first bit must always be a one. This theoretically allows for eight message types, but there are some special cases in the form of system messages (see below).

Standard MIDI messages can be up to three bytes long, but not all messages require three bytes, and there are some fairly common exceptions to the rule which are described below. Table 8.1 shows the format and content of the main MIDI message types, but they will not be described in any greater detail here.

8.4 An overview of software for MIDI

Sequencers are probably the most ubiquitous of the available MIDI software packages. A sequencer will be capable of storing a number of 'tracks' of MIDI information, editing it and otherwise manipulating it for musical composition purposes. It is also capable of storing MIDI events for non-musical purposes such as studio automation, and may be equipped with digital audio recording capabilities in some cases. Some of the more advanced packages are available in modular form (allowing the user to buy only the functional blocks required) and in cut-down or 'entry-level' versions for the new user. Sequencer software is discussed in greater detail below.

The dividing line between sequencer and music notation software is a grey one, since there are features common to each. Music notation software is designed to allow the user control over the detailed appearance of the printed musical page, rather as desktop publishing packages work for typesetters, and such software often provides facilities for MIDI input and output. MIDI input is used for entering note pitches during setting, whilst output is used for playing the finished score in an audible form. Most major packages will read and write standard MIDI files, and can therefore exchange data with sequencers, allowing sequenced music to be exported to a notation package for fine tuning of printed appearance. It is also common for sequencer packages to offer varying degrees of music notation capability, although the scores which result are rarely as professional in appearance as those produced by dedicated notation software.

Librarian and editor software is used for managing large amounts of voice data for MIDI-controlled instruments. Such packages communicate with MIDI instruments using system exclusive messages in order to exchange parameters relating to voice programs. The software may then allow these voice programs or 'patches' to be modified using an editor, offering a rather better graphical interface than those usually found on the front panels of most sound modules. Banks of patches may be stored on disk by the librarian, in order that libraries of sounds can be managed easily, and this is often cheaper than storing patches in the various 'memory cards' offered by synth manufacturers. Banks of patch information may be accessed by sequencer software in order that the operator may choose voices for particular tracks by name, rather than by program change numbers. Sample editors are also available, offering similar facilities, although sample dumps using system exclusive are not really recommended, unless they are short, since the time taken can be excessive. Sample data can be transferred to a computer using a faster interface than MIDI (such as SCSI) and the sample waveforms can be edited graphically.

Amongst other miscellaneous software packages available for the computer are MIDI mixer automation systems, guitar sequencers, MIDI file players, multimedia authoring systems and alternative user interfaces. Development software is also available for MIDI programmers, providing a programming environment for the writing of new MIDI software applications. There are also a number of software applications designed principally for research purposes or for experimental music composition.

8.5 Interfacing a computer to a MIDI system

In order to use a computer as a central controller for a MIDI system it must have at least one MIDI interface, consisting of at least an IN and an OUT port. (THRU is not strictly necessary in most cases.) Unless the computer has a built-in interface, as found on old Atari machines, some form of third-party hardware interface must be added, and there are many ranging from simple single ports to complex multiple port products.

8.5.1 Single port MIDI interfaces

A typical single port MIDI interface will be connected either to one of the spare I/O ports of the computer, or plugged into an expansion slot. On the Macintosh, for example, a MIDI interface is normally connected directly to one of the two serial ports (which are multi-purpose RS-422 type interfaces capable of operating at rates up to a few hundred kilobits per second), as shown in Figure 8.9(a). Any software then asks the user to configure the serial ports to suit the interface, requiring him or her to choose which of the two serial ports will be used and what the notional master clock speed of the UART will be (Figure 8.9(b)). (The choice of clock speeds is something of a hang-over from earlier days, when a 1 or 2 MHz clock speed was the master clock in contemporary computers.) Serial data arriving at the interface's MIDI IN is then converted into a pseudo RS-422 electrical format and fed to the computer's serial port to be treated like any other external serial data. Outgoing serial data is converted into the MIDI current loop electrical form.

MIDI interfaces for the PC have taken a somewhat different development line. The built-in DOS and BIOS routines for the PC limited serial communications to 9600 baud, requiring MIDI software writers to bypass these routines and write directly to the serial ports. The hardware specification for PC serial ports should allow them to be operated at a high enough rate for MIDI, but this does not always turn out to be the case in practice. For these reasons, and because of the relatively coarse timing

Figure 8.9 (a) A spare general purpose serial interface such as the Macintosh modem port can be used to connect an external MIDI interface

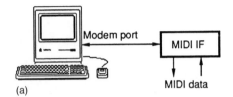

(a)

(b) The interface is configured in software, using a utility such as that shown here

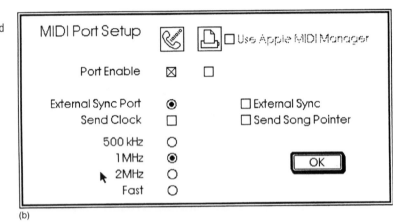

(b)

facilities available in the PC operating system, the Japanese company Roland developed an interface for the PC called the MPU-401 which itself contained a microprocessor whose job was to handle many of the MIDI processing tasks concerned with sending and receiving note data and synchronisation information. The interface ICs were licensed to other developers, and consequently the MPU-401-compatible interface has become something of a standard for the PC. Most PC software will be able to address it, although it is truly only a single port interface even when it sports more than one MIDI OUT (all the ports carry the same information).

MPU-401 compatible interfaces connect directly to a selection of lines on the data, control and address buses of the PC, and use CPU interrupts to transfer MIDI data to and from the PC's memory. The MPU interface occupies either a memory or I/O mapped port address which may be addressed as port 0 (or anything from 0 to 3 depending on the number of interfaces). The processor on the interface acts as a 'background processor', allowing the computer's main CPU to handle its normal tasks of screen update, user interface communications and disk storage without undue delays, whilst storing or playing MIDI data. Such an interface will also handle simple synchronisation using the older timing signals of drum click and FSK. The original MPU-

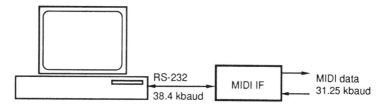

Figure 8.10 A PC may not be able to run its RS-232 serial interface at 31.25 kbaud, so the MIDI interface performs the function of rate translation to and from the higher 38.4 kbaud rate of the computer

401 is now no longer available from Roland, and has been replaced by the MPU-IPC. There is also an IBM PS/2 expansion card available. A number of other manufacturers also make MPU-401 compatible interfaces.

Modern PCs have serial interfaces that will operate at a high enough data rate for MIDI, but are not normally able to operate at precisely the 31.25 kbaud required, thus rendering them virtually unusable for direct translation into the MIDI electrical format. None the less, there are a few external interfaces available which connect to the PC's serial port and transpose a higher serial data rate (often 38.4 kbaud) down to the MIDI rate using intermediate buffering and flow control (see Figure 8.10). These would be addressed differently by MIDI software, compared with the MPU-401, requiring MIDI data to be routed directly to the serial port rather than writing it to an internal expansion port.

An alternative approach with the PC is to install a non-MPU-compatible card which drives a MIDI interface. This would require suitable driver software, capable of addressing the specific interface.

8.5.2 Multiport interfaces

The MIDI protocol only allows 16 receiver channels to be addressed directly. Provided that electrical conditions were properly controlled it would be possible to connect a number of receiving devices to a single port but the limit of 16 channels would remain. For this reason, multiport interfaces have become widely used in MIDI systems where more than 16 channels are required, and they are also useful as a means of limiting the amount of data which is sent or received through any one MIDI port. (A single port can become 'overloaded' with MIDI data if serving a large number of devices, resulting in data delays.)

An example of a simple approach to multiple port MIDI interfacing is seen on the Apple Macintosh (Mac), since the Mac has two almost identical serial ports (the printer port and the modem

Figure 8.11 The Macintosh computer has two serial ports, each of which may be connected to an independent MIDI interface. The effect is to double the number of addressable MIDI channels provided that the software is capable of sending data to both ports

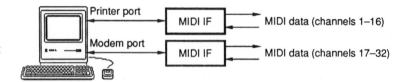

port). In such cases a single port MIDI interface could be connected to each of the serial ports, as shown in Figure 8.11, thereby doubling the number of channels which might be addressed. This requires software capable of addressing 16 channels via each of the independent ports, and most of the advanced packages can do this. Some require the user to decide which of the two ports is to be used for MIDI. Clearly this is only an option if the serial ports are both available. If one of them is required for other purposes then it may be possible to obtain an interface with a serial 'THRU' switch, which allows the serial data from say the printer port to be connected either to the MIDI interface or looped through to another serial port which could be connected to a serial peripheral such as a printer or modem (see Figure 8.12). Quite often, one of the Mac's serial ports is used for networking, and this renders that port unavailable for MIDI.

Figure 8.12 A serial 'thru' switch is sometimes provided on an external MIDI interface so that the serial data can be daisy-chained on to a non-MIDI serial device such as a printer. This saves repeated re-plugging of cables

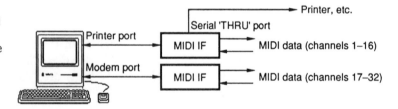

A more advanced approach involves the use of an external MIDI interface which has a number of independent MIDI OUT ports (each with its own UART). Such an interface is connected to the host computer using either a parallel or serial I/O port, or using an expansion card, as shown in Figure 8.13. The principle of such approaches is that data is transferred between the computer and the multiport interface at a higher speed than the normal MIDI rate (called 'Fast Mode' on the Mac), requiring the interface's CPU to distribute the MIDI data between the output ports as appropriate, and transmit it at the normal MIDI rate. This requires a certain amount of 'intelligence' in the interface, and the routing of data is normally performed under the control of the host computer which will configure the MIDI interface according to the studio setup designed by the user.

Figure 8.13 A multiport MIDI interface is connected to the computer using either a serial or parallel interface running a number of times faster than the MIDI data rate

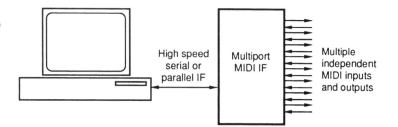

Most multiport systems allow up to 16 channels to be addressed by each of the MIDI OUTs, thus expanding the total number of channels to 16 times the number of ports. In such systems it is common for each instrument to be connected to its own port, both IN and OUT, which is especially useful with multi-timbral sound modules capable of operating on all 16 channels simultaneously. It also allows any instrument to be used as a controller, and makes it possible for instruments to send system exclusive information back to the computer. Multiport interfaces will often allow data received from more than one port to be merged, or allow recording from more than one source at a time.

An interesting commercial example of a multiport MIDI interface exists in the form of Opcode's Studio 5, pictured in Figure 8.14. It has two serial ports which transfer data to and from the Mac, shown in Figure 8.15, and these can be 'THRU'd' if required to connect other serial devices (although the Studio 5 then loses this port for MIDI communications). Serial communication between the Studio 5 and the computer normally operates at a higher rate than MIDI, and the two serial ports provide a greater data throughput capacity than just one. The control software shares the data between the two serial ports in order to optimise flow, and it is therefore not possible to say that one

Figure 8.14 Front panel of the Opcode Studio 5 multiport MIDI interface showing two MIDI ports for 'guest' instruments, and LEDs to show transmit and receive activity on each port

Figure 8.15 Rear panel of the Opcode Studio 5, showing Macintosh serial ports and serial THRU ports, as well as MIDI interfaces, timecode interface and other external connections

serial port serves only certain MIDI ports. There are 15 MIDI ports with two on the front panel for easy connection of 'guest' devices or controllers that are not installed at the back, and the software which controls the Studio 5 allows the user to configure the studio so that the computer 'knows' which instrument is connected to which port. Applications then address instruments rather than ports or channels. The Studio 5 also incorporates other facilities for manipulating and routing MIDI data, and it has timecode ports in order that timecode information can be relayed to and from the computer in the form of MTC (MIDI TimeCode).

8.5.3 Interface driver software

Quite commonly, a software driver is provided for an external MIDI interface or expansion card. This driver is used by the computer's operating system to address the MIDI interface concerned, and it takes care of handling the various routines necessary to carry data to and from the physical I/O ports. The MIDI application therefore simply talks to the driver, and it follows that you must have the correct driver installed for the interface which you intend to use. High-end MIDI software usually comes complete with a number of drivers and configuration documents for the most popular MIDI interfaces.

8.6 MIDI operating systems

In order to manage the MIDI data which is used by software applications, some manufacturers have designed operating system extensions which enhance the system capabilities of a computer. Such extensions typically operate in the background and are unobtrusive in normal operation. Features vary, but can include keeping a map of the MIDI devices connected to a multiport interface, performing filtering and processing of MIDI data for certain input and output ports, synchronising multiple MIDI applications and optimising the throughput of MIDI data to ensure optimum timing accuracy.

An example of such an approach is Opcode's OMS (Opcode MIDI System). OMS acts as a system integrator mainly for Opcode's software products, but it has been adopted by other manufacturers including Apple for QuickTime. OMS allows the user to configure the studio setup in software to represent the physical MIDI connections which exist. With a multiport interface the layout of the connections is configured in a window which might look something like the diagram in Figure 8.16.

Figure 8.16 Example of an OMS studio setup document showing which devices are connected to each MIDI port

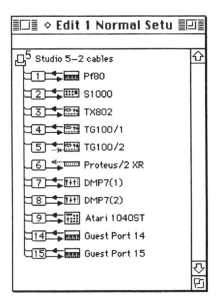

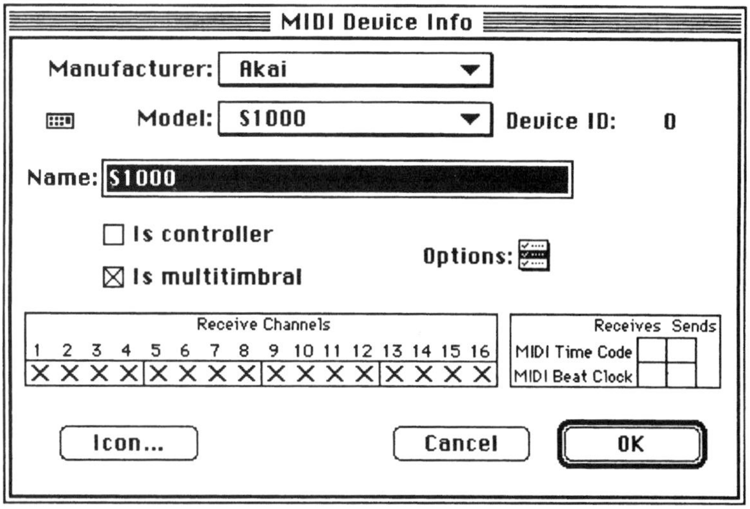

Figure 8.17 Each device in the OMS setup is described in such a window as this. Here the user determines what MIDI information is received and transmitted by the device, and sets the SysEx device ID

Here the user has defined instruments which are connected to each physical port, and each instrument's definition includes information about the type of data it receives and transmits, as shown in Figure 8.17. It is possible to indicate which MIDI channels an instrument will receive on, and whether or not it transmits and receives synchronisation data.

Once a studio setup has been defined in this manner, all OMS compatible applications have access to the information. In a sequencer package, for example, the user may select an instrument for a particular track by name, and it is also possible to select which instrument(s) will act as sources for MIDI data to be recorded. It is no longer necessary to worry about which physical ports they are connected to. If the studio setup is changed then all OMS applications may be updated with the new conditions automatically.

OMS also allows for applications to be run in a synchronised fashion, by forming 'soft links' between the timing clocks of the applications. In this way two separate sequencers could be made to run as if they were one, or a digital audio editing package could be locked to a sequencer, provided that they were both OMS compatible.

8.7 Introduction to sequencer concepts

The section that follows is intended to provide an overview of some of the most common concepts associated with MIDI sequencers. This is an introductory explanation, showing how MIDI is implemented in sequencers, and the concepts explained here may be considered as 'core' concepts which apply almost no matter what the product or which the revision of the software. Readers who wish to find more detailed coverage of sequencer implementations and their applications are recommended to refer to Yavelow's excellent *Music and Sound Bible* as listed at the end of this chapter. Examples are taken from commercial packages, to illustrate specific points, but clearly individual packages will differ in matters of terminology and detail. A lot of the examples illustrated here are taken from Opcode software, but similar concepts are found in other manufacturers' products.

8.7.1 Tracks and channels

A sequencer may be presented to the user so that it emulates to some extent the controls of a multitrack tape recorder. The example shown in Figure 8.18 illustrates this point, showing the familiar transport controls in the top left hand corner. There are advantages in this approach, since it is a familiar interface for

Figure 8.18 Example of transport controls in a sequencer window (Opcode Studio Vision)

many users, but some clarification is required regarding the way in which the 'tracks' of this MIDI recorder can be made to relate to MIDI channels, as it is possible, for example, to encounter sequencers with many more tracks than there are MIDI channels.

A track is simply a way of presenting memory space to the user, and it also helps in the compartmentalisation of information so that particular sections of material may be assigned to particular tracks. There is also the conceptual advantage that a composition or recording is built up by successively overlaying more and more tracks, all of which may be played together at the same time. Unlike the tape recorder, the MIDI recorder's tracks are not fixed in their time relationship and can be slipped against each other, as they simply consist of data stored in the memory. On older or less advanced sequencers, the replay of each track is assigned to a particular MIDI channel, and this results in data from that track being output with its status bytes set to that channel number. It may not matter what channel was operative when the track was recorded, as the sequencer allows the user to change it on replay, perhaps even during the song. More recent packages are considerably more flexible in this respect, offering an almost unlimited number of virtual tracks and allowing tracks to be assigned to virtual instruments. A track can contain data for more than one channel. Using a multiport MIDI interface it is possible to address a much larger number of instruments than the basic 16 channels of MIDI allowed in the past, and operating systems such as OMS allow these instruments to be chosen by name rather than by channel or port number.

8.7.2 Input and output filters

As MIDI information is received from the hardware interface(s) it will be stored in memory, and, unless anything is done to prevent it, all the data that arrives will be stored. If memory space is limited it may be helpful to filter out some information before it can be stored, using an input filter. This will be a sub-section of the program which watches out for the presence of certain MIDI status bytes and their associated data as they arrive, so that they can be discarded before storage. The user may be able to select input filters for such data as aftertouch, pitch bend, control changes and velocity information, among others. Clearly it is only advisable to use input filters if it is envisaged that this data will never be needed, since although filtering saves memory space the information is lost for ever. Filtering may also help to speed up the MIDI system on replay, owing to the reduced data flow, although it may slow down the computer during recording.

If memory space were not limited, it would be possible to store all the data that arrived and filter it at the output if it became necessary to prevent certain data from being transmitted on replay. Output filters are often implemented for similar groups of MIDI messages as for the input filters. It may also be possible to filter out timing and synchronisation data for those devices which do not require it. Some input and output filtering may also be performed in the MIDI operating system or environment, in which it is often possible to determine which instruments transmit and receive particular types of MIDI data, in order to limit the data flow over individual MIDI cables to a minimum.

8.7.3 Displaying and editing MIDI information

A sequencer is the ideal tool for manipulating MIDI information, and this may be performed in a number of ways depending on the type of interface provided to the user. The most flexible is the graphical interface employed on many desktop computers which may provide for visual editing of the stored information either as a musical score, a table or event list of MIDI data, or in the form of a grid of some kind. Figure 8.19 shows a number of examples of different approaches to the display of stored MIDI information. Although it might be imagined that the musical score would be the best way of visualising MIDI data, it is often not the most appropriate. This is partly because unless the input is successfully quantised (see below) the score will represent precisely what was played when the music was recorded, and this is rarely good looking on a score! The appearance is often messy because some parts were slightly out of time, and the literal interpretation of what was played may be almost unrecognisable compared with what you thought you played. Score representation is useful after careful editing and quantisation, and can be used when you need to produce a visually satisfactory printed output. Alternatively, you might prefer to save the score as a MIDI file and export it to a music notation package for layout purposes.

In the grid editing display, notes may be dragged around using a mouse or trackball, and audible feedback is often available as the note is dragged up and down, allowing the user to hear the pitch or sound as the position changes. Note lengths may be changed by dragging their ends in or out, and the timing position may be altered by dragging the note left or right. In the event list form, each MIDI event is listed next to a time value. The information in the list may then be changed by typing in

Figure 8.19 MIDI data, once stored in sequencer memory, can be displayed in a variety of forms. These are some examples: (a) event list (Studio Vision); (b) graphical form (Studio Vision); (c) musical score (Notator Logic)

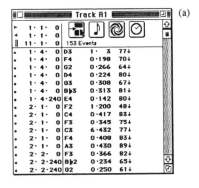

(a)

(b)

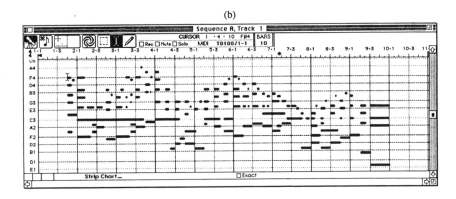

(c)

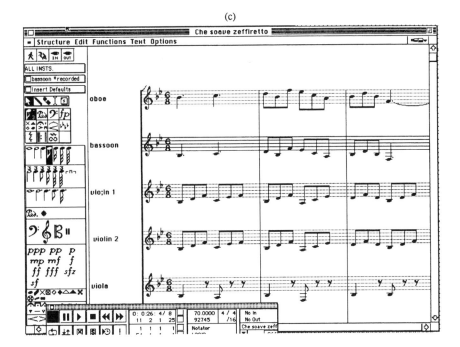

new times or new data values. Also events may be inserted and deleted. In all of these modes the familiar cut and paste techniques used in word processors and other software can be applied, allowing events to be used more than once in different places, repeated so many times over, and other such operations.

A whole range of semi-automatic editing functions are also possible, such as transposition of music, using the computer to operate on the data so as to modify it in a predetermined fashion before sending it out again. Transposition, for example, is simply a matter of raising or lowering the MIDI note numbers of every stored note by the relevant degree. You could also create echo effects by duplicating a track and offsetting it by a certain amount, for example. A sequencer's ability to search the stored data (both music and control) based on specific criteria, and to perform modifications or transformations to just the data which matches the search criteria, is one of the most powerful features of a modern system. For example, it may be possible to search for the highest-pitched notes of a polyphonic track so that they can be separated off to another track as a melody line. Alternatively it may be possible to apply the rhythm values of one track to the pitch values of another so as to create a new track, or to apply certain algorithmic manipulations to stored durations or pitches for compositional experimentation.

8.7.4 Quantisation of rhythm

Rhythmic quantisation is a feature of almost all sequencers, and is based upon the sequencer's ability to alter the timing of events. In its simplest form it involves the 'pulling-in' of note events to the nearest musical time interval at the resolution specified by the user, so that notes which were 'out of time' can be played back 'in time' (see Figure 8.20). It is normal to be able to program the quantisation resolution to an accuracy of at least as small as a 32nd note, and the choice depends on the audible effect desired. It must be borne in mind that notes played far enough out of time that they are over the halfway mark between one time interval and the next will go to the nearest time interval, which might be the one before or the one after that intended. In this case, editing would be required.

Events may be quantised either permanently or just on replay. The permanent method will alter the timing for ever after, whereas the replay-only quantisation may be changed at will and does not affect the stored data. Some systems allow 'record quantisation' which alters the timing of events as they arrive at the input to the sequencer. This is a form of permanent quantisation. It may also

Figure 8.20 In simple rhythm quantisation, notes (represented by X) are 'pulled in' to the nearest musical timing point (the minimum increment is represented by *t*). Notes over the halfway mark between one timing point and the next will be quantised to the next point. Advanced algorithms are available which allow the user to decide within what range out-of-time notes will be 'pulled in', and others make it possible to leave alone those notes which are only slightly out of time and quantise those more drastically adrift, thereby preserving the natural irregularity in playing whilst correcting obvious mistakes

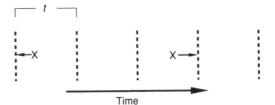

be possible to 'quantise' the cursor movement so that it can only drag events to predefined rhythmic divisions.

More complex rhythmic quantisation is also possible, in order to maintain the 'natural' feel of rhythm for example. Simple quantisation can result in music which sounds 'mechanical' and electronically produced, whereas the 'human feel' algorithms available in many packages attempt to quantise the rhythm strictly and then reapply some controlled randomness. The parameters of this process may be open to adjustment until the desired effect is achieved. Figure 8.21 shows an example of the controls provided by one package to control the nature of quantisation.

Figure 8.21 Quantisation options in Opcode Studio Vision. 'Sensitivity' determines how close a note has to be to the quantising interval (grid) to be quantised. If negative then further away notes are quantised and closer notes are left alone, whereas positive values have the opposite effect. 'Strength' determines how far notes are pulled towards quantising intervals; 'Shift' quantises notes and then shifts them out of time by so many units; 'Swing' moves every other quantising interval so that it is not exactly in between the two on either side; 'Smear' randomly moves notes after they have been quantised. If the duration is quantised then note durations are adjusted to the nearest specified increment

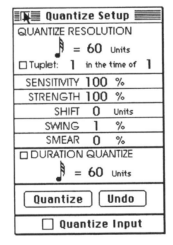

8.7.5 Non-note events in MIDI sequences

In addition to note events, one may either have recorded or may wish to add events which are for other control purposes, such as program change messages, controller messages or system exclusive messages. If such messages are transmitted from the

Figure 8.22 Two examples
of control information
displayed in a sequencer's
strip chart. (a) Note
velocities, and (b) continuous
controller data (e.g. MIDI
volume). (Opcode EZ Vision)

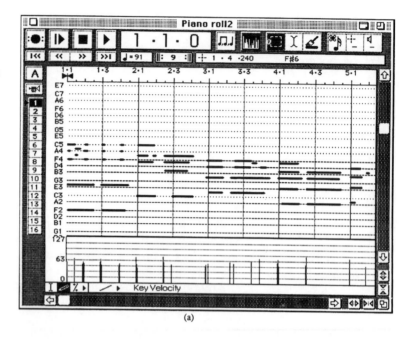

(a)

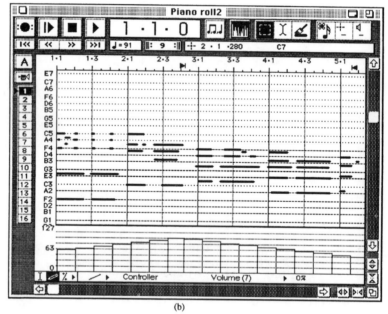

(b)

MIDI data source during recording then they will normally be
stored as events alongside the note data (unless they have been
filtered out). Such data may be displayed in a number of ways,
but again the graphical plot is arguably the most useful. It is

common to allow selected controller data to be plotted below the note data in a strip chart, such as shown in Figure 8.22. Here two examples are shown. The first one shows note velocities and the second shows movements of the volume controller.

It is possible to edit these events in a similar way to note events, but there are a number of other possibilities here. For example a scaling factor may be applied to controller data in order to change the overall effect by so many per cent, or a graphical contour may be drawn over the controller information to scale it according to the magnitude of the contour at any point. Such a contour could be used to introduce a gradual increase in note velocities over a section, or to introduce any other time-varying effect.

System exclusive data may also be recorded or inserted into sequences in a similar way to the message types described above. Any such data received during recording will normally be stored and may be displayed in a list form. It is also possible to insert SysEx voice dumps into sequences in order that a device may be loaded with new parameters whilst a song is executing if required.

8.7.6 Digital audio in MIDI sequencers

Many of the leading sequencer software packages optionally offer digital audio capabilities. Typically, such software requires the presence of one of the common third party audio expansion cards or suitable internal audio processing capability within the workstation. For example, nearly all the Macintosh-based sequencers which offer digital audio require the installation of at least one Digidesign audio card such as 'ProTools', 'Sound Tools' or 'AudioMedia'. It is very convenient operationally when common digital audio hardware is used by multiple software packages, since it makes possible easy switching between applications and only requires the purchase of one audio expansion kit.

Integrated MIDI and digital audio information is presented to the user in a number of different ways, and an example is shown in Figure 8.23. Tracks are usually depicted horizontally, either as a detailed waveform display of the audio signal which may be zoomed in to various levels of detail to allow precise editing, or as named blocks occupying a certain duration. The audio editing features are often cut down equivalents to those found in dedicated audio editing packages.

When integrated with MIDI sequencing it may be possible to quantise digital audio events rhythmically, rather as MIDI

Figure 8.23 MIDI and digital audio data may be displayed and edited alongside each other, as shown in this example from Digidesign's ProTools package

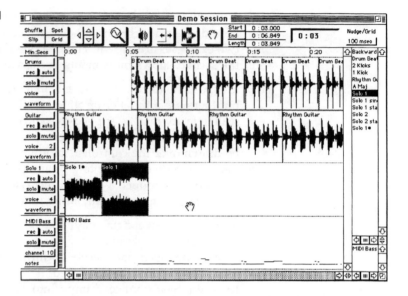

events can be quantised. This is achieved in one package by allowing the user to strip out the silence between audio events (say the beats of a drum) by removing everything below a certain audio threshold. Thereafter each beat of the drum is a separate audio 'entity' and its start point may be rhythmically quantised. One package also has a remarkable function which takes a melodic line from a digital audio file and converts it into equivalent MIDI note information, complete with velocity, volume and brightness controller information, allowing it to be edited like any other MIDI data. Controller data such as pitch bend information can then be modified and subsequently reapplied to the audio data, allowing degrees of audio manipulation not previously possible. Many of the other arrangement features found in sequencers can also be applied to the digital audio tracks. In this way, MIDI control data and digital audio data may be edited and arranged in parallel.

8.8 Standard MIDI files

Sequencers and notation packages typically store data on disk in their own unique file formats. Occasionally it is possible for a file from one package to be read by others, especially if the packages are from the same manufacturer, but this is rare. The standard MIDI file was developed in an attempt to make interchange of information between packages more straightforward, and is now used widely in the industry in addition to manufacturers' own file formats.

Three types of standard MIDI file exist to encourage the interchange of sequencer data between software packages. The MIDI file contains data representing events on individual sequencer tracks, as well as containing labels such as track names, instrument names and time signatures. It is the intention that not only should software packages running on the same computer type be able to read these universal files, but that such files may be ported to other computers (using one of the many file exchange protocols), so that a package running under a completely different operating system may read the data.

File type 0 is the simplest, and is used for single-track data, whilst file type 1 supports multiple tracks which are 'vertically' synchronous with each other (such as the parts of a song), and file type 2 contains multiple tracks which have no direct timing relationship and may thus be asynchronous. Type 2 could be used for transferring song files which are made up of a number of discrete sequences, each with a multiple track structure.

The basic file format consists of a number of 8 bit words formed into so-called 'chunks', rather like the IFF format described in Chapter 6. The header chunk, which always heads a MIDI file, contains global information relating to the whole file, whilst subsequent track chunks contain event data and labels relating to individual sequencer tracks. Track data should be distinguished from MIDI channel data, since a sequencer track may address more than one MIDI channel. Each chunk is preceded by a preamble of its own, which specifies the type of chunk (header or track) and the length of the chunk in terms of the number of data bytes which are to be contained in the chunk. There then follows the designated number of data bytes (see Figure 8.24).

Figure 8.24 The general format of a MIDI file chunk. Each chunk has a preamble consisting of a 4 byte ASCII 'type' followed by 4 bytes to represent the number of data bytes in the rest of the message (the 'length')

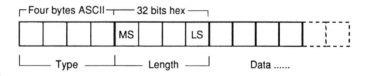

As can be seen, the chunk preamble contains first a 4 byte section to identify the chunk type using the ASCII character format (four bytes equivalent to four characters), and second a 4 byte section (eight hexadecimal characters) to indicate the number of data bytes in the chunk (the length). The number of bytes indicated in this length message does not include the length of the preamble (which is always eight bytes).

Figure 8.25 The header chunk has the type 'MThd' and the number of data bytes indicated in the 'length' is 6 (see text)

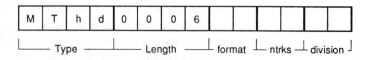

The header chunk which begins every MIDI file takes the format shown in Figure 8.25. After the 8 byte preamble will normally be found 6 bytes containing header data, considered as three 16 bit words, the first of which ('format') defines the file type as 0, 1 or 2 (see above), the second of which ('ntrks') defines the number of track chunks in the file, and the third of which ('division') defines the timing format used in subsequent track events. A zero in the MSB of the 'division' word indicates that events will be represented by 'musical' time increments of a certain number of 'ticks per quarter note' (the exact number is defined in the remaining bits of the word), whilst a one in the MSB indicates that events will be represented by real-time increments in number-of-ticks-per-timecode-frame. The frame rate of the timecode is given in the remaining bits of the most significant byte of 'division', being represented using negative values in two's complement form. Thus the standard frame rates are represented by one of the decimal values –24, –25, –29 (for 30 drop frame) or –30. A negative number may be represented in two's complement form by taking its positive binary equivalent, inverting all the bits (zeros become ones and vice versa), then adding a binary one.

When a real-time format is specified (MSB of 'division' = 1) in the header chunk, the least significant byte of 'division' is used to specify the subdivisions of a frame to which events may be timed. For example, a value of '4_{10}' in this position would mean that events were timed to an accuracy of a quarter of a frame, corresponding to the arrival frequency of MIDI quarter-frame timecode messages, whilst a value of '80_{10}' would allow events to be timed to bit accuracy within the timecode frame (there are 80 bits representing a single timecode frame value in the SMPTE/EBU longitudinal timecode format).

Following the header come a number of track chunks (see Figure 8.26), the number depending on the file type and the number of tracks. File type 0 represents a single track, and thus will only contain a header and one track chunk, whilst file types 1 and 2 may have many track chunks. Track chunks contain strings of MIDI events, each labelled with a delta-time at which the event is to occur. Delta-times represent the number of 'ticks' since the last event, as opposed to the absolute time since the beginning of a song. The exact time increment specified by a tick depends

Figure 8.26 A track chunk has the type 'MTrk' and the number of data bytes indicated in the 'length' depends on the contents of the chunk. The data bytes which follow are grouped into events as shown

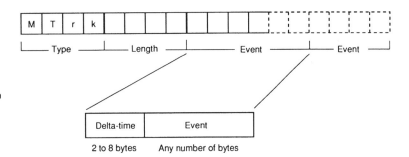

on the definition of a tick contained in the 'division' word of the header (see above).

Delta-time values are represented in 'variable length format', which is a means of representing hexadecimal numbers up to &0FFFFFFF as compactly as possible. Variable length values represent the number in question using one, two, three or four bytes, depending on the size of the number. Each byte of the variable length value has its MSB set to a one, except for the last byte whose MSB should be zero. (This distinguishes the last byte of the value from the others, so that the computer reading the data knows when to stop compiling the number.) There are thus seven bits of each byte available for the representation of numeric data (rather like the MIDI status and data bytes). A software routine must be written to convert normal hex values into this format and back again.

8.9 General MIDI

If MIDI files are to be exchanged between systems and replayed on different sound generating hardware, one needs a means of ensuring that the music will sound at least similar on the two systems. One of the problems with MIDI-controlled sound generators used to be that although voice programs could be selected using MIDI program change commands, there was no standard mapping of voices to program numbers. In other words, program change 3 might correspond to 'alto sax' on one instrument and 'grand piano' on another. Consequently a music sequence could sound completely different when replayed on two different multitimbral sound generators. General MIDI was introduced to standardise some basic elements of sound generator control, so that MIDI files could be exchanged more easily between systems, ensuring that an approximation to the sounds and instruments intended by the composer would be heard no matter what General MIDI sound generator was used for replay.

Currently, General MIDI is specified at Level 1, although there are proposals to extend the concept to further levels.

General MIDI specifies a number of other things as well as standard sounds. For example, it specifies a minimum degree of polyphony, and requires that a sound generator should be able to receive MIDI data on all 16 channels simultaneously and polyphonically, with a different voice on each channel. There is also a requirement that the sound generator should support percussion sounds in the form of drum kits, so that a General MIDI sound module is capable of acting as a complete 'band in a box'. Some multimedia computers now incorporate General MIDI synthesisers as part of the sound hardware, and have operating systems capable of addressing MIDI instruments.

Dynamic voice allocation is the norm in GM sound modules, with a requirement either for at least 24 dynamically allocated voices in total, or 16 for melody and 8 for percussion. In order to ensure compatibility between sequences that are replayed on GM modules, percussion sounds are always allocated to MIDI channel 10. Program change numbers are mapped to specific voice names, with ranges of numbers allocated to certain types of sounds, as shown in Table 8.2. Precise voice names may be found in the GM documentation (see the end of the chapter). Channel 10, the percussion channel, has a defined set of note numbers on which particular sounds are to occur, so that the composer may know for example that key 39 will always be a 'hand clap'.

Table 8.2 General MIDI program number ranges (except channel 10)

Program change (decimal)	Sound type
0–7	Piano
8–15	Chromatic percussion
16–23	Organ
24–31	Guitar
32–39	Bass
40–47	Strings
48–55	Ensemble
56–63	Brass
64–71	Reed
72–79	Pipe
80–87	Synth lead
88–95	Synth pad
96–103	Synth effects
104–111	Ethnic
112–119	Percussive
121–128	Sound effects

8.10 MIDI TimeCode (MTC)

MIDI Timecode (MTC) is used widely as a means of synchronising MIDI-controlled equipment to a real time reference. A number of multiport MIDI interfaces have a SMPTE/EBU timecode port which may be used to read and write longitudinal timecode (LTC) from and to recording equipment, and this is often converted to MTC so that it can be handled by associated computer software. It is common for audio workstation software to use MTC as a sync reference, especially when digital audio is combined with MIDI.

In an LTC timecode frame (see section 7.6), two binary data groups are allocated to each of hours, minutes, seconds and frames, these groups representing the tens and units of each, so there are eight binary groups in total representing the time value of a frame. In order to transmit this information over MIDI, it has to be turned into a format which is compatible with other MIDI data (i.e. a status byte followed by relevant data bytes). There are two types of MTC synchronising message: one which updates a receiver regularly with running timecode, and another which transmits one-time updates of the timecode position for situations such as exist during the high speed spooling of tape machines, where regular updating of each single frame would involve too great a rate of transmitted data. The former is known as a quarter-frame message, denoted by the status byte (&F1), whilst the latter is known as a full-frame message and is transmitted as a universal realtime system exclusive (SysEx) message, the details of which will not be given here.

One timecode frame is represented by too much information to be sent in a standard three byte MIDI message, so it is broken down into eight separate messages called quarter-frame messages. Four of these quarter frame messages are transmitted in the period of one timecode frame (in order to limit the message rate), so it takes two frames to transmit a complete frame value and the receiver is updated every two frames. Receivers therefore need to maintain a two frame offset between their displayed timecode and the last decoded value, because by the time a frame value has been completely transmitted two frames will have elapsed. Internally, the timing resolution of software can be made higher than that of the timecode messages, with the messages being used as a reference point every so often.

Each message of the group of eight represents a part of the timecode frame value, as shown in Figure 8.27, and takes the general form:

&[F1] [DATA]

Figure 8.27 General format of the quarter-frame MTC message

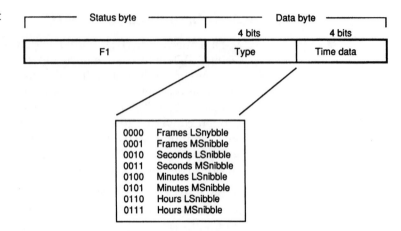

0000	Frames LSnybble
0001	Frames MSnibble
0010	Seconds LSnibble
0011	Seconds MSnibble
0100	Minutes LSnibble
0101	Minutes MSnibble
0110	Hours LSnibble
0111	Hours MSnibble

The data byte begins with zero (as always), and the next seven bits of the data word are made up of a 3 bit code defining whether the message represents hours, minutes, seconds or frames, MSnibble or LSnibble, followed by the four bits representing the binary value of that nibble. In order to reassemble the correct timecode value from the eight quarter-frame messages, the LS and MS nibbles of hours, minutes, seconds and frames are each paired within the receiver to form 8 bit words.

8.11 MIDI Machine Control (MMC)

MIDI may be used for remotely controlling tape machines and other studio equipment. MMC (MIDI Machine Control) uses universal realtime SysEx messages, and has a lot in common with a remote control protocol known as 'ESbus' which was devised by the EBU and SMPTE as a universal standard for the remote control of tape machines, VTRs and other studio equipment. The ESbus standard uses an RS422 remote control bus running at 38.4 kbaud, whereas the MMC standard uses the MIDI bus for similar commands. Although MMC and ESbus are not the same, and the message protocols are not identical, the command types and reporting capabilities required of machines are very similar.

There are a number of levels of complexity at which MMC can be made to operate, and communication is possible in both closed and open loop modes (either with or without a return connection to the controlling computer). By allowing this flexibility, MMC makes it possible for designers to implement it at anything from a very simple level (i.e. cheaply) to a very complicated level involving all the finer points. MMC is gaining increasing popularity in semi-professional equipment, because it is somewhat cheaper to implement than ESbus and allows equipment to be integrated easily with a MIDI-based studio system. There are a number of tape machines and synchronisers now on the market with MIDI interfaces, and some sequencer packages handle the remote control of studio machines using the MMC protocol. Studio machines may be connected to the main studio computer by connecting them to one port on a multiport MIDI interface.

MMC could be used simply to control the basic transport functions of an audio tape recorder. In such a case only a very limited set of commands would need to be implemented in the tape recorder, and very little would be needed in the way of responses. In fact it would be quite feasible to operate the transport of a tape recorder using an open-loop approach – simply sending 'play', 'stop', 'rewind', etc. as required by the controlling application.

8.12 Troubleshooting a MIDI system

When a MIDI system fails to perform as expected, or when devices appear not to be responding to data which is being transmitted from a controller, it is important to adopt logical fault-finding techniques rather than pressing every button in sight and starting to replug cables. The fault will normally be a simple one, and there is only a limited number of possible causes. It is often worth starting at the end of the system nearest to the device which exhibits the problem and working backwards towards the controller, asking a number of questions as you go. You are basically trying to find out either where the control signal is getting lost or why the device is responding in a strange way. The old saying: 'If it aint broke, don't try to fix it' is a good one. Many computer-controlled MIDI systems stay working properly because of a cocktail of good luck and perseverance with sorting out a combination of software and hardware that works together.

8.12.1 Device not responding?

Look at the hints in Figure 8.28. First, is MIDI data getting to the device in question? Most devices have some means of indicating that they are receiving MIDI data, either by a flashing light on the front panel or some other form of display. Alternatively it is possible to buy small analysers which in their simplest form may do something like flashing a light if MIDI data is received. If data is getting to the device then the problem is probably either within the device or after its audio output. The most

Figure 8.28 In this diagram are shown a number of suggestions to be considered when troubleshooting a MIDI system

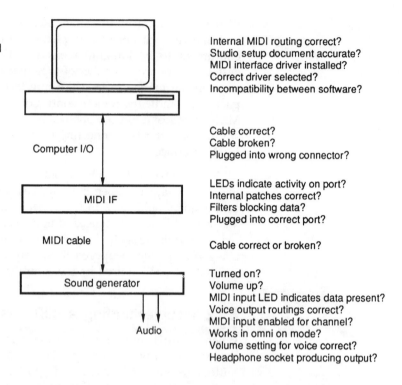

Computer I/O

Internal MIDI routing correct?
Studio setup document accurate?
MIDI interface driver installed?
Correct driver selected?
Incompatibility between software?

Cable correct?
Cable broken?
Plugged into wrong connector?

MIDI IF

LEDs indicate activity on port?
Internal patches correct?
Filters blocking data?
Plugged into correct port?

MIDI cable

Cable correct or broken?

Sound generator

Audio

Turned on?
Volume up?
MIDI input LED indicates data present?
Voice output routings correct?
MIDI input enabled for channel?
Works in omni on mode?
Volume setting for voice correct?
Headphone socket producing output?

common mistake that people make is to think that they have a MIDI problem when in fact they have an audio problem. Check that the audio output is actually connected to something and that its destination is turned on and faded up. Plug in a pair of headphones to check if the device is responding to MIDI data. If sound comes out of the headphones then the problem most probably lies in the audio system.

If the device is receiving MIDI data but not producing an audio output, try setting the receive mode to 'omni on' so that it responds on all channels. If this works then the problem must be related to the way in which a particular channel's data is being handled. Check that the device is enabled to receive on the MIDI channel in question. Check that the volume is set to something other than zero, and that any external MIDI controllers assigned to volume are not forcing the volume to zero (such as any virtual faders in the sequencer package). Check that the voice assigned to the channel in question is actually assigned to an audio output which is connected to the outside world. Check that the main audio output control on the unit itself is turned up. Also try sending note messages for a number of different notes – it may be that the voice in question is not set up to respond over the whole note range.

If no MIDI data is reaching the device then move one step further back down the MIDI signal chain. Check the MIDI cable. Swap it for another one. If the device is connected to a MIDI patcher or router of some kind, check that the patcher input receiving the required MIDI data is routed to the output concerned. Try connecting a MIDI keyboard directly to the patcher input concerned to see if the patch is working. If this works then the problem lies further up the chain, either in the MIDI interface attached to the controller or in the controller itself. If the controller is a computer with an external MIDI interface, it may be possible to test the MIDI port concerned. The setup software for the MIDI interface may allow you to enter a 'Test' mode in which you can send unspecified note data directly to the physical port concerned. This should test whether or not the MIDI interface is working. Most interfaces have lights to show when a particular port is receiving or transmitting data, and this can be used for test purposes. It may be that the interface needs to be reconfigured to match a changed studio setup. Now go back to the controller and make sure that you are sending data to the right output on the required MIDI channel and that you are satisfied, from what you know about it, that the software concerned should be transmitting.

If no data is getting from the computer to the interface, check the cables to the interface. Then try resetting the interface and the computer. This sometimes re-establishes communication between the two. Reset the interface first, then the computer, so that the computer 'sees' the interface (this may involve powering down, then up). Alternatively, a soft reset may be possible using the setup software for the interface. If this does not work, check that no applications are open on the computer which might be taking over the interface ports concerned (some applications will not give up control over particular I/O ports easily). Check the configuration of any software MIDI routers within the computer to make sure that MIDI data is 'connected' from the controlling package to the I/O port in question.

Ask yourself the question: 'was it working the last time I tried it?'. If it was, it is unlikely that the problem is due to more fundamental reasons such as the wrong port drivers being installed in the system or a specific incompatibility between hardware and software, but it is worth thinking through what you have done to the system configuration since the last time it was used. It is possible that new software extensions or new applications may conflict with your previously working configuration, and removing them will solve the problem. Try using a different software package to control the device which is not responding. If this works then the problem is clearly with the original package.

8.12.2 Device responding in an unexpected way

Assuming that the device in question had been responding correctly on a previous occasion, any change in response to MIDI messages such as program and control changes is most likely due either to an altered internal setup or a message getting to the device which was not intended for it.

Most of the internal setup parameters on a MIDI-controlled device are accessible either using the front panel or using system exclusive messages. It is often quite a long-winded process to get to the parameter in question using the limited front panel displays of many devices, but it may be necessary to do this in order to check the intended response to particular MIDI data. If the problem is one with unusual responses (or no response) to program change messages then it may be that the program change map has been altered, and that a different stored voice or patch is being selected from the one intended. Perhaps the program change number in question is not assigned to a stored voice or patch at all. If the device is switching between programs when it should not then it may be that your MIDI routing is at fault. Perhaps the device is receiving program changes intended for another. Check the configuration of your MIDI patcher or multiport interface. A similar process applies to controller messages. Check the internal mapping of controller messages to parameters, and check the external MIDI routing to make sure that devices are receiving only the information intended for them.

When more than one person uses a MIDI-controlled studio, or when you have a lot of different setups yourself, virtually the only way to ensure that you can reset the studio quickly to a particular state is to store system exclusive dumps of the full configuration of each device, and to store any patcher or MIDI operating system maps. These can either be kept in separate librarian files or as part of a sequence, to be downloaded to the devices before starting the session. Once you have set up a configuration of a device which works for a particular purpose it should be stored on the computer so that it could be dumped back down again at a later date.

8.13 Sound sampling and synthesis hardware for desktop computers

A large number of sound cards are available for desktop computers, capable of acting as multitimbral (many voices at one time) sound generators, with stereo outputs. Alternatively an increasing number of multimedia computers now have such sound generating capacity built in as part of the standard hardware. This sound generating hardware is often controlled using the internal equivalent of MIDI control, even though the sound hardware may not actually be connected using a standard MIDI interface. General MIDI, as described above, is becoming the standard way of ensuring that MIDI sound files replay in a manner which makes the music sound similar no matter which computer system or sound card is used. Both Microsoft's *Windows* and Apple's Macintosh system software now include General-MIDI-like approaches to synthesised sound generation for multimedia applications. As introduced at the start of the chapter, synthesised sound generation under MIDI control is far more economical on storage space and processing power than digital sound file replay, and may be acceptable for many applications in which it is not necessary to replay the *exact* musical sound which was recorded, but rather an approximation to it.

The two main approaches to synthetic sound generation in PCs are FM and wavetable synthesis. In FM synthesis, as pioneered by John Chowning and developed by Yamaha, the frequency of one oscillator (or 'operator') is modulated by another oscillator or chain of oscillators, as shown in Figure 8.29. The result of this frequency modulation (FM) is the creation of a complex set of sidebands or spectral components around the fundamental or 'carrier' frequency of the last oscillator in the chain, as exemplified in Figure 8.30. Quite rich timbres can be created with only a few oscillators/operators, although advanced FM synthesisers often use up to 6 operators per voice. The operators can be arranged in different ways, either in a chain with each modulating the next, or partly in parallel with each sub-chain contributing a particular component of the voice. These configurations are

Figure 8.29 In FM synthesis one operator (the equivalent of an oscillator) frequency modulates another so as to alter its output spectrum. Each operator has its own envelope generator which affects how the output level of the operator changes with time

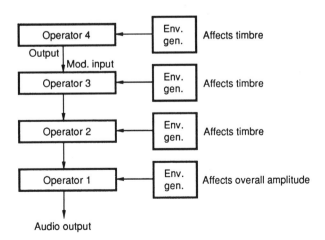

273

Figure 8.30 The result of FM is the creation of a sideband pattern around the 'carrier' oscillator frequency (f_c). The amplitudes and frequencies of these sidebands depend on the amplitude and frequency of the modulating signal (f_m). Sidebands are spaced apart by the frequency of the modulating signal, and a higher modulator amplitude generally creates more sidebands (a richer timbre). Sidebands which fall into the negative frequency range (below 0 Hz) are folded back into the positive range, some with phase reversal

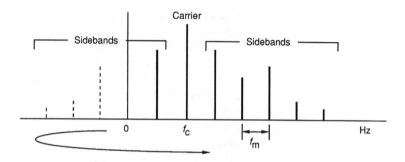

called 'algorithms', and some examples are shown in Figure 8.31. Each operator can be affected by an envelope generator which controls the way in which the amplitude of the output changes with time, and a simple envelope has 4 stages, as shown in Figure 8.32.

Although FM is a very flexible way of producing new synthesised sounds it is not always easy to predict or program, so as to produce a particular desired output. Wavetable synthesis is more predictable in this respect, since it involves the storage of short portions of sampled sound waves in memory (the wavetable is basically the series of memory addresses containing the discrete

Figure 8.31 Some examples of FM synthesis algorithms. (a) Operators in parallel give the equivalent of additive synthesis (not really FM). (b) Operators in series produce very complex and unpredictable timbres. (c) A combination of series and parallel operators can be used so that different components of the sound can be handled by different parts of the algorithm

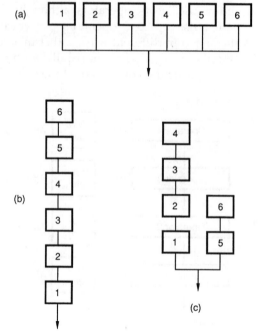

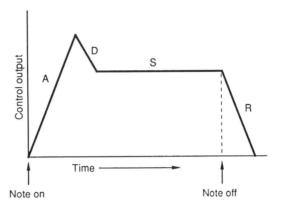

Figure 8.32 A typical envelope generator has four stages. Attack (A), Decay (D), Sustain (S) and Release (R). The rates and maximum values of each stage of the envelope can be set independently

sample values of the stored wave segment). During replay, the wave samples are read out of memory in various ways, very similar to the replay of ordinary digital audio recordings, except that the pitch of the stored sound is varied by skipping samples in order to change the period of the replayed sound. Using variable rate replay and digital filtering (techniques akin to sample rate conversion and pitch shifting, as described in Chapter 2) a simple stored wave segment can be transformed in both pitch and timbre. A technique known as looping is used to allow quite short stored wave segments to be lengthened or sustained on replay by repeating one section of the stored wave over and over. There is a clear trade off here between the shortness of the looped segment (which conserves memory) and the quality of the instrumental sound produced. As with other forms of synthesis, envelope generators are used to alter the characteristics of the output over the duration of each note, and often separate wavetables are used for attack and sustain portions of a note.

Wavetables stored in ROM are inflexible because only a standard sound set is available to the multimedia sound designer. Increasingly common is for sound cards to contain an area of working RAM, in addition to standard sounds on ROM, which can be used for the storage of temporary or custom wavetables. In this way the sound designer can ensure that the sounds he wants to use are available when his product is played, by arranging for the appropriate audio samples to be uploaded into the sound card RAM from hard disk or CD-ROM before the action begins. New sounds could be loaded during the course of a game or other multimedia production.

In addition to sound synthesis hardware for multimedia PCs, it is also possible to obtain full sound sampling hardware and software. Such systems allow the user to record and edit his or her own sound samples, store them on disk and replay them polyphonically from RAM in the same way as a stand-alone MIDI-controlled sampler. Since sound cards of this type are usually connected directly to the computer's expansion bus, it is normal to control them by establishing a 'soft' connection to appropriate sequencer software using the computer's MIDI control system extensions such as OMS (described above). The sound card then behaves like a MIDI-controlled device but is actually communicating over the computer's expansion bus.

Recommended further reading

Heckroth, J. (1993) *A Tutorial on MIDI and Wavetable Synthesis* . Crystal Semiconductors Application Note AN27. Crystal Semiconductors, PO Box 17847, Austin, TX 78760, USA.

MMA (1983) *MIDI 1.0 Detailed Specification.* MIDI Manufacturers Association.

MMA (1986) *MIDI Sample Dump Standard.* MIDI Manufacturers Association.

MMA (1987) *MIDI Timecode and Cueing: Detailed Specification.* MIDI Manufacturers Association.

MMA (1988) *Standard MIDI Files 1.0.* MIDI Manufacturers Association.

MMA (1991) *General MIDI System Level 1.* MIDI Manufacturers Association.

MMA (1991) *MIDI Show Control 1.0.* MIDI Manufacturers Association.

MMA (1992) *MIDI Machine Control 1.0.* MIDI Manufacturers Association.

MMA (1993) *4.2 Addendum to MIDI 1.0 Specification.* MIDI Manufacturers Association.

Rumsey, F. (1994) *MIDI Systems and Control,* 2nd edition, Focal Press.

Yavelow, C. (1987) 'Computers and music: the state of the art'. *Journal of the Audio Engineering Society* , **35**, 3, pp. 161–193.

Yavelow, C. (1992) *Macworld Music and Sound Bible.* IDG Books Worldwide, Inc., San Mateo, CA., USA.

The MIDI Manufacturers Association can be contacted by email at: *mma@earthlink.net* or on WorldWide Web at *http://www.earthlink.net/~mma.*

Index